FOCUS ON SUBTRACTION

K-6 CONTINUITY MATHEMATICS

The enclosed material is designed for educational purposes only. Each State may have different certification and specific guidelines. Please refer to your State for additional and future information. The information contained herein is considered correct at the time of creation but laws and regulations are updated frequently and the reader assumes the responsibility for confirming current regulations and applicable data. The publisher and author make no warranty as to the success of the individuals using the training material contained herein. The publisher and author make no warranty as to any action taken by any individual completing this program. The reader is responsible for the appropriate use of the materials and information provided. This publication is designed to provide accurate and authoritative information concerning the subject matter. All material is sold with the understanding that neither the author nor the publisher guarantees the actions of any individual making use of the inclusions. Neither the author nor the publisher is rendering a legal opinion, accounting recommendation or other professional service. If legal advice or other expert assistance is desired, the services of a legal professional or other individual should be sought. The applicable federally released forms, disclosures and notices are generated from public domain. Copyright law does apply to all intellectual materials and all rights under said law are reserved b y the copyright owner.

Coursework is available at special quantity discounts to use as premiums and sales promotions within corporate or private training programs. To obtain information or inquire about availability please write to Director, PO Box 1, Hollidaysburg, PA 16648.

FOCUS ON SUBTRACTION

K-6 CONTINUITY MATHEMATICS

Note to Teaching Helper:

The first years of school are transitional years taking a child from a play approach to an academic approach to learning. As you move through the foundations of mathematics, the instruction will become more structured and contain many new concepts and facts for your student to master. It is important that your mathematics lessons encompass a variety of delivery methods. The best programs will encompass both practical worksheets and hands on activities that encourage a deeper understanding of the ideas behind the computations necessary for success in mathematics.

The worksheets included in the skill mastery textbook should serve as an overall guide to the mathematics program and a tool to help you determine your student's mastery of each necessary math building block. You should encourage hands on, creative approach to mathematics as a method of fixing the necessary concepts in your students mind and encouraging a true understanding of the processes of mathematics.

Many learners enjoy using concrete objects or tools to help them master mathematics concepts. You should work with your student to decide what concrete tools work best with their learning style. You might choose any group of pre-made math manipulative, common household items, tally marks, or almost any item that illustrates a specific sum of objects to help your student on the path to a true understanding of the building blocks of mathematics.

The important factor to remember as you move through the each level of the mathematics continuity program is that mastery of each new skill exceeds the need for completion of the entire program during a pre-set term. Statistics have shown that a child will succeed at mathematics only if they gain comprehension of each new concept. You should use the included guide and worksheets to assess your child's progress and only move on to the next segment once you are certain that mastery has been achieved.

The proven path to mastery in mathematics is practice, practice, practice. This does not mean that you should present mathematics in the form of mindless repetition. You should present varied opportunities to approach each mathematics concept from a variety of angles. You should also encourage reinforcement of each previously mastered concept along with the presentation of each new concept. This allows the child to build upon a solid foundation of comprehension that will encourage a true mastery of mathematics throughout their educational career.

The focus on series of lessons bring a laser focus to each subject area, ensuring each necessary skill is fully developed before the student moves onto the next skill in the series.

Continuity is critical to educational success. By ensuring each student gains a comprehensive understanding of each skill before transitioning to the next skill, you lay the groundwork for lifelong academic success.

Unit 1 – Counting to Subtract

Instructor Note

Your student should gain a basic understanding of the concept that subtraction decreases the final number.

1.   X the box that shows 1 less sun.

2.   X the box that shows 1 less ball.

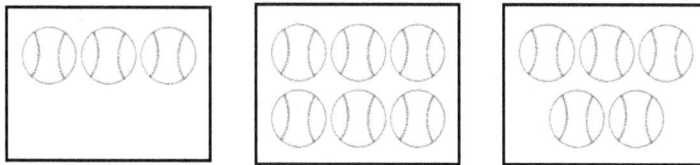

3. X the box that shows 1 less crayon.

4. X the box that shows 1 less apple.

5. X the box that shows 1 less frog.

Name: _____

1. X the box that shows 2 less elephants.

2. X the box that shows 2 less fish.

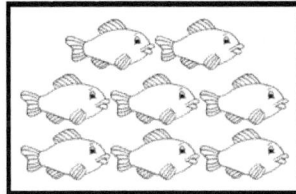

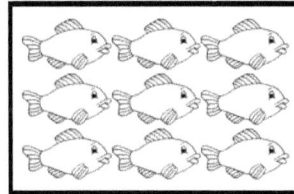

3. X the box that shows 2 less flags.

4.   X the box that shows 2 less pigs.

5.   X the box that shows 2 less butterflies.

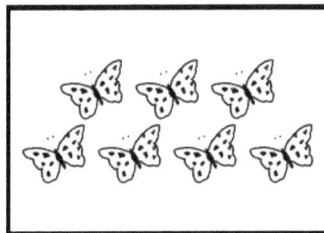

Name: _____

1.   X the box that shows 3 less hats.

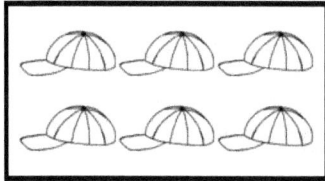

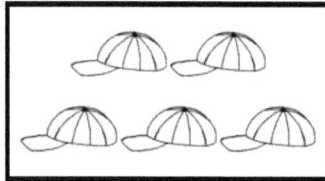

2.   X the box that shows 3 less birds.

3.   X the box that shows 3 less triangles.

4. X the box that shows 3 less alligators.

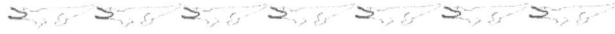

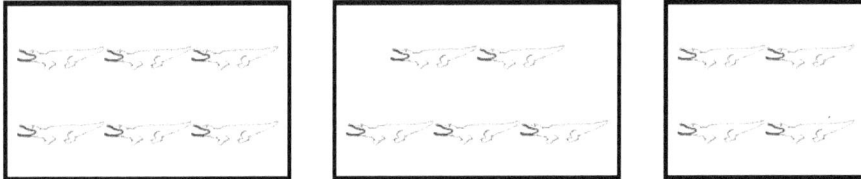

5. X the box that shows 3 less jets.

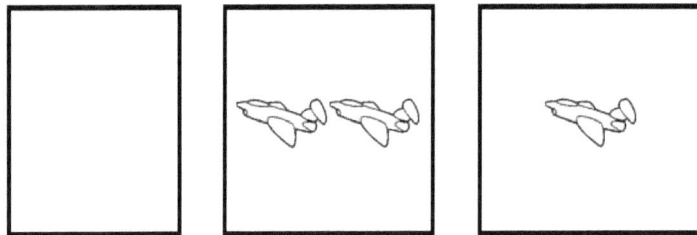

6. X the box that shows 3 less cakes.

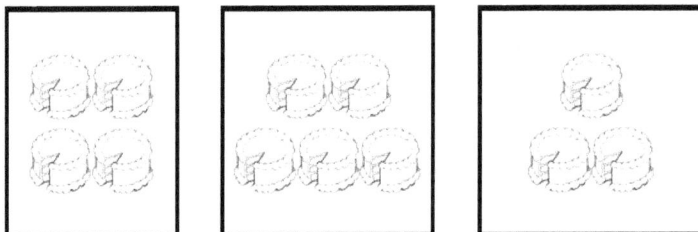

Name: _____

1.   Cross out 4 balls. Write how many balls you have now. _____

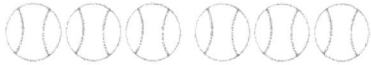

⬤⬤⬤⬤ ⬤⬤⬤⬤

2.   Cross out 4 triangles. Write how many triangles you have now. _____

△△△ △△△ △△△

3.   Cross out 4 arrows. Write how many arrows you have now. _____

⬆⬆⬆⬆⬆ ⬆ ⬆⬆⬆⬆

4.   Cross out 4 circles. Write how many 4circles you have now. _____

◯◯◯◯◯◯◯◯◯◯

5.   Cross out 4 hats. Write how many hats you have now. _____

Name: _____

1. Cross out 5 cans. Write how many cans you have now. _____

2. Cross out 5 cats. Write how many cats you have now. _____

3. Cross out 5 drums. Write how many drums you have now. _____

4. Cross out 5 flowers. Write how many flowers you have now. _____

5. Cross out 5 rectangles. Write how many rectangles you have now.
_____

Name: _____

1.   Cross out 6 nuts. Write how many nuts you have now. _____

2.   Cross out 6 pans. Write how many pans you have now. _____

3.   Cross out 6 squares. Write how many squares you have now. _____

4.   Cross out 6 trees. Write how many trees you have now. _____

5.   Cross out 6 tubs. Write how many tubs you have now. _____

Unit 2 – Model Subtraction

Instructor Note:

Your student should gain the ability to use models to apply basic subtraction facts.

Student Instructions:

When you subtract one group of objects from another group of objects, they make a smaller group.

If you have a group of objects and take away some of the objects, you are subtracting. How many are left is a clue that you need to subtract.

6 Stars	3 Stars	3 Stars Left

1.   Subtract the drums and then X the box that shows the right answer.

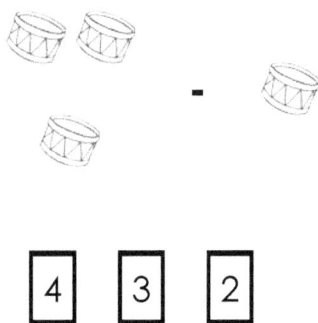

| 4 | 3 | 2 |

2.   Subtract the bats and then X the box that shows the right answer.

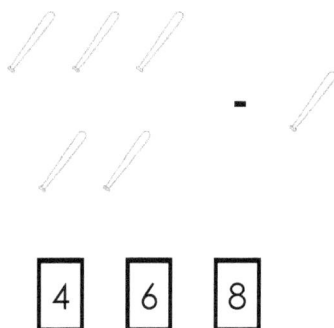

| 4 | 6 | 8 |

3. Subtract the cans and then X the box that shows the right answer.

| 4 | 3 | 1 |

4. Subtract the cats and then X the box that shows the right answer.

| 7 | 6 | 8 |

5. Subtract the hens and then X the box that shows the right answer.

| 4 | 3 | 5 |

13

Name: _____

1. Subtract the circles and then X the box that shows the right answer.

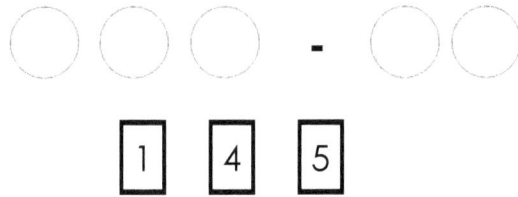

○ ○ ○ - ○ ○

| 1 | | 4 | | 5 |

2. Subtract the clowns and then X the box that shows the right answer.

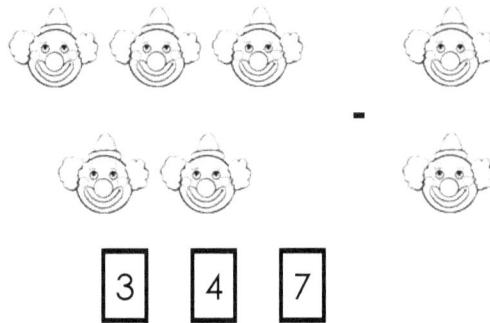

| 3 | | 4 | | 7 |

3. Subtract the crabs and then X the box that shows the right answer.

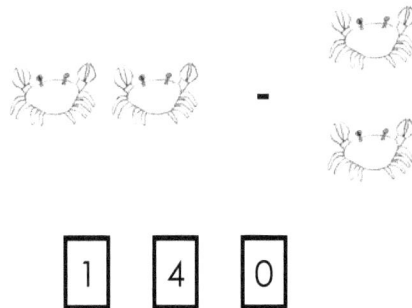

| 1 | | 4 | | 0 |

4. Subtract the dogs and then X the box that shows the right answer.

| 5 | 6 | 9 |

---

5. Subtract the elephants and then X the box that shows the right answer.

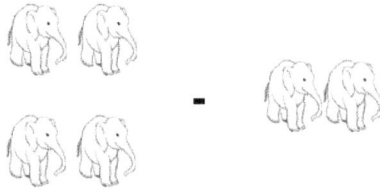

| 2 | 4 | 6 |

15

Name: _____

1. Subtract the fish and then X the box that shows the right answer.

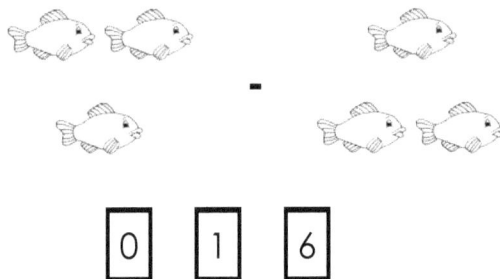

| 0 | 1 | 6 |

2. Subtract the flags and then X the box that shows the right answer.

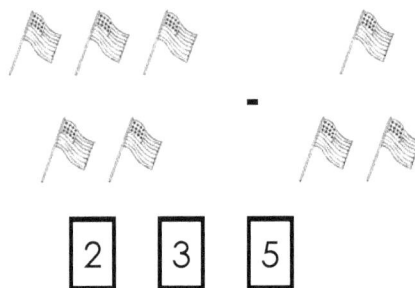

| 2 | 3 | 5 |

3. Subtract the flowers and then X the box that shows the right answer.

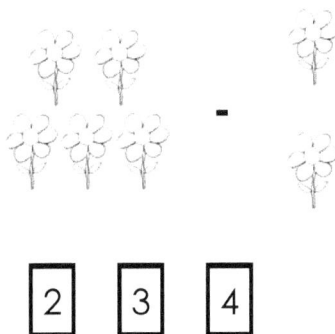

| 2 | 3 | 4 |

4. Subtract the gifts and then X the box that shows the right answer.

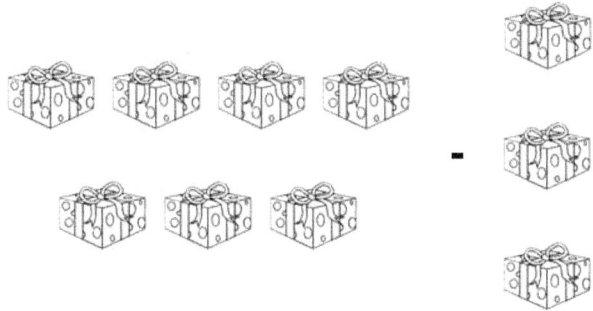

| 4 | 10 | 5 |

---

5. Subtract the hats and then X the box that shows the right answer.

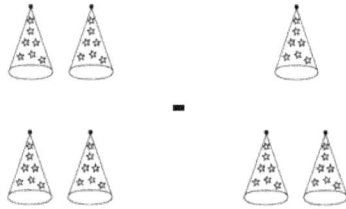

| 1 | 0 | 7 |

17

1. Subtract the pigs and then X the box that shows the right answer.

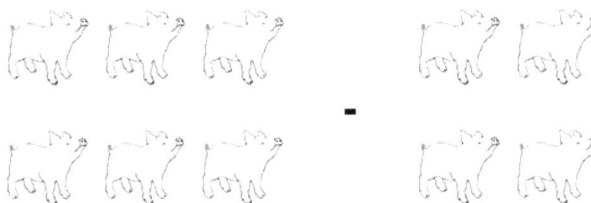

| 1 | | 2 | | 3 |

2. Subtract the penguins and then X the box that shows the right answer.

| 1 | | 2 | | 3 |

3. Subtract the porcupines and then X the box that shows the right answer.

| 2 | | 3 | | 4 |

4. Subtract the beds and then X the box that shows the right answer.

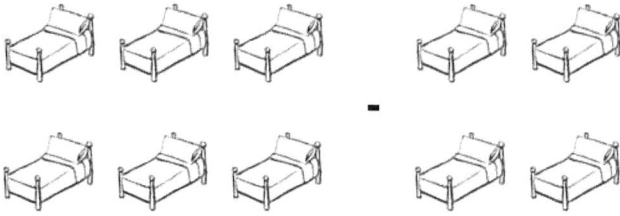

▪

| 3 | | 2 | | 1 |

---

5. Subtract the eels and then X the box that shows the right answer.

▪

| 2 | | 1 | | 0 |

---

1. Subtract the lions and then X the box that shows the right answer.

| 1 | 2 | 3 |

2. Subtract the nuts and then X the box that shows the right answer.

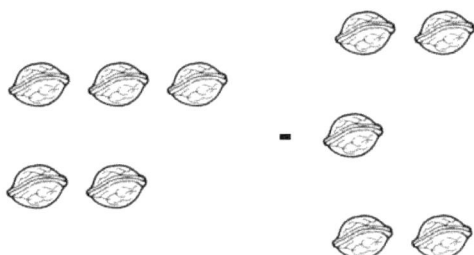

| 0 | 1 | 2 |

3. Subtract the nets and then X the box that shows the right answer.

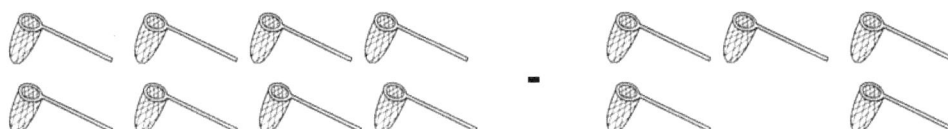

| 3 | 2 | 1 |

4. Subtract the pans and then X the box that shows the right answer.

| 1 | | 2 | | 3 |

---

5. Subtract the rats and then X the box that shows the right answer.

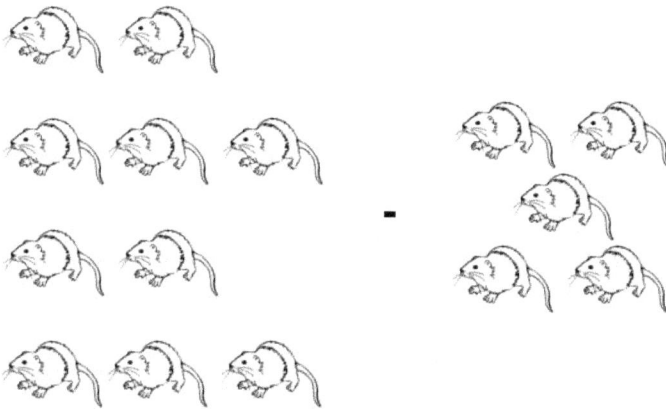

| 7 | | 6 | | 5 |

21

Name: _____

1. Subtract the rectangles and then X the box that shows the right answer.

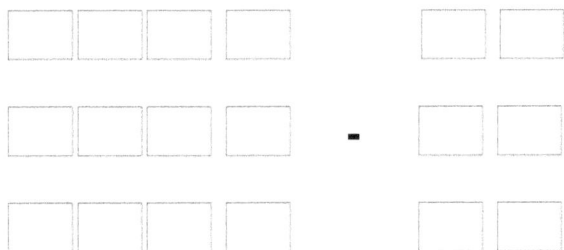

[ ] [ ] [ ] [ ]   [ ] [ ]

[ ] [ ] [ ] [ ]   -   [ ] [ ]

[ ] [ ] [ ] [ ]   [ ] [ ]

| 7 | | 6 | | 5 |

2. Subtract the trains and then X the box that shows the right answer.

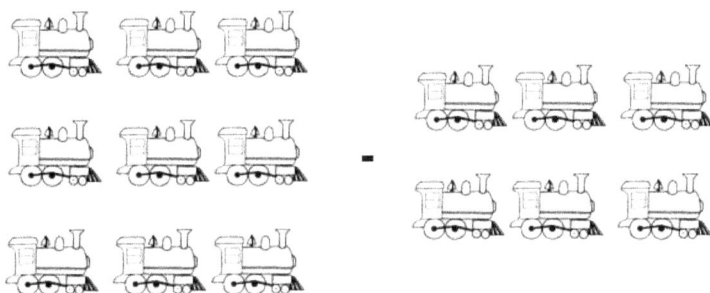

| 2 | | 3 | | 4 |

3. Subtract the bins and then X the box that shows the right answer.

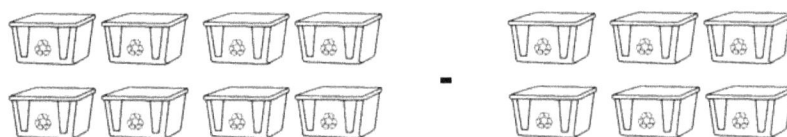

| 3 | | 2 | | 4 |

4. Subtract the hands and then X the box that shows the right answer.

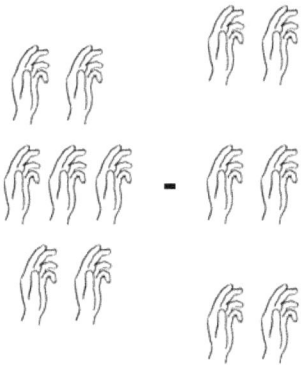

| 3 | 2 | 1 |

---

5. Subtract the sharks and then X the box that shows the right answer.

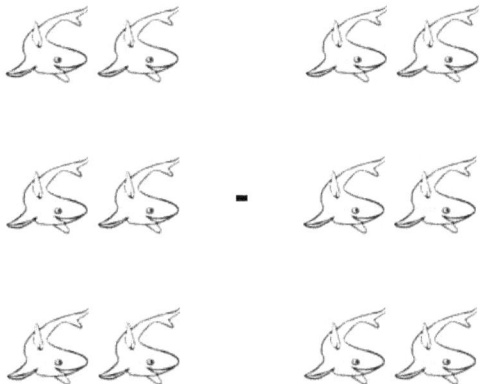

| 0 | 1 | 2 |

Name: _____

1. Subtract the goats and then X the box that shows the right answer.

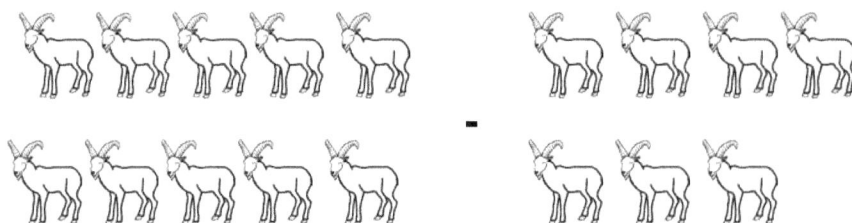

| 4 | 3 | 2 |

2. Subtract the squares and then X the box that shows the right answer.

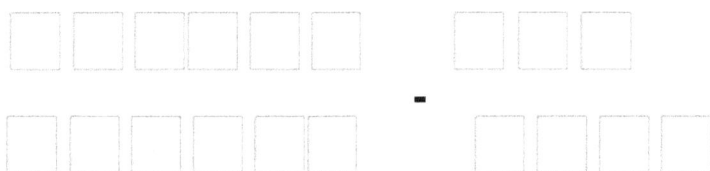

| 4 | 5 | 6 |

3. Subtract the pens and then X the box that shows the right answer.

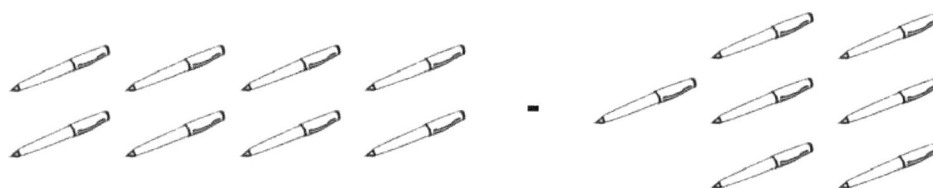

| 2 | 1 | 0 |

24

4. Subtract the tents and then X the box that shows the right answer.

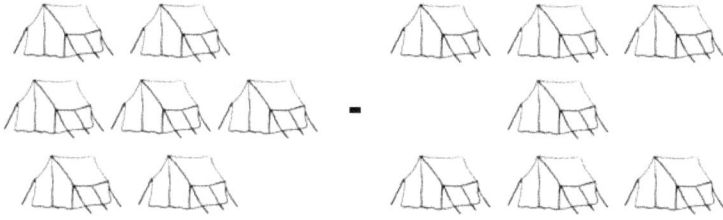

| 0 | 1 | 2 |

---

5. Subtract the bears and then X the box that shows the right answer.

| 1 | 3 | 2 |

Name: _____

1. Subtract the trees and then X the box that shows the right answer.

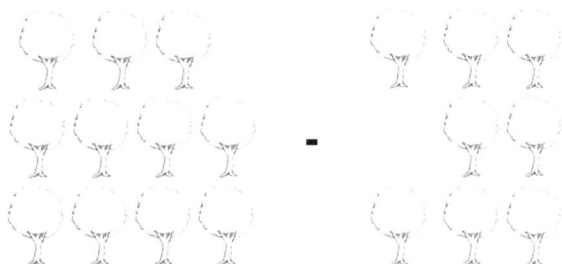

| 4 | 3 | 2 |

2. Subtract the whales and then X the box that shows the right answer.

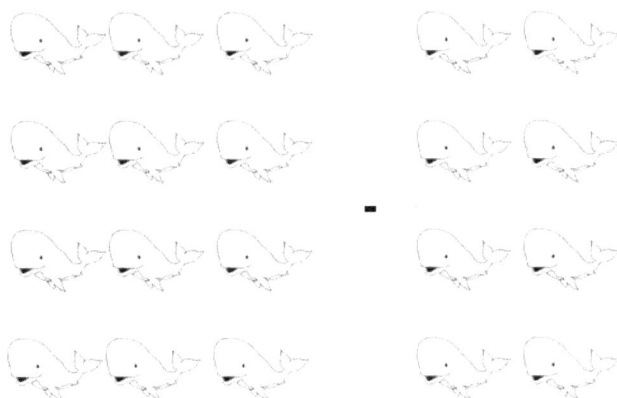

| 3 | 4 | 5 |

3. Subtract the logs and then X the box that shows the right answer.

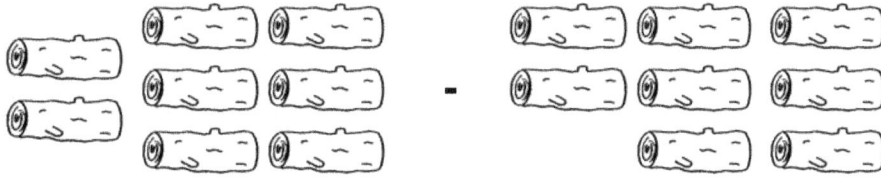

| 2 | | 1 | | 0 |

---

4. Subtract the pins and then X the box that shows the right answer.

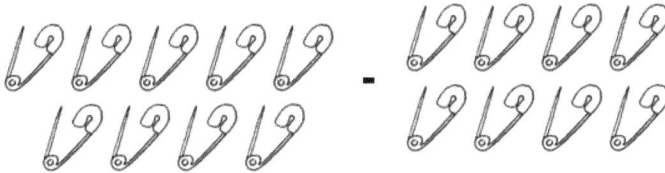

| 0 | | 1 | | 2 |

---

5. Subtract the tubs and then X the box that shows the right answer.

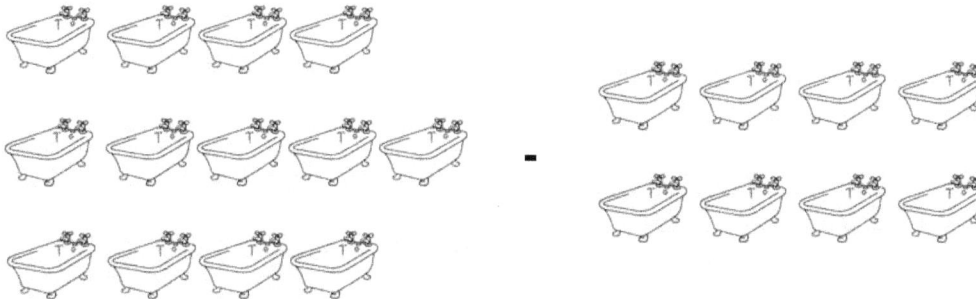

| 4 | | 5 | | 6 |

Name: _____

1. Subtract the cakes and then X the box that shows the right answer.

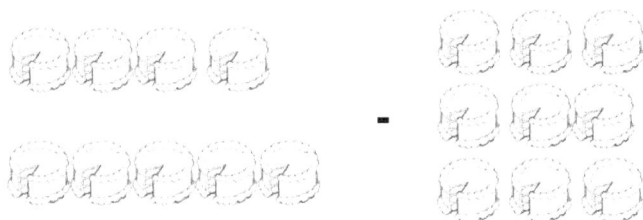

| 2 | 1 | 0 |

2. Subtract the triangles and then X the box that shows the right answer.

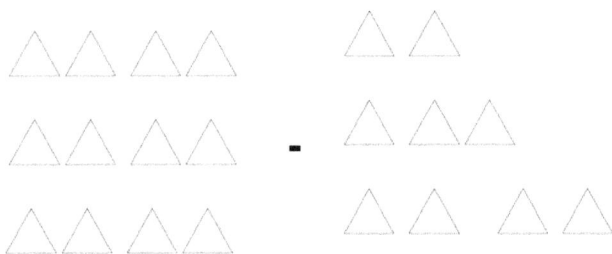

| 1 | 3 | 2 |

3. Subtract the fish and then X the box that shows the right answer.

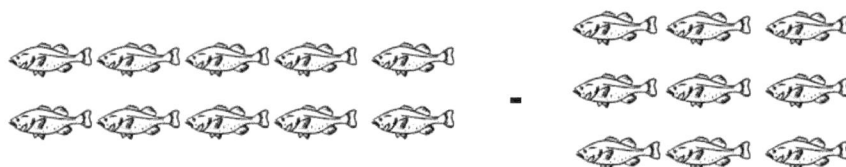

| 1 | 2 | 3 |

4. Subtract the balls and then X the box that shows the right answer.

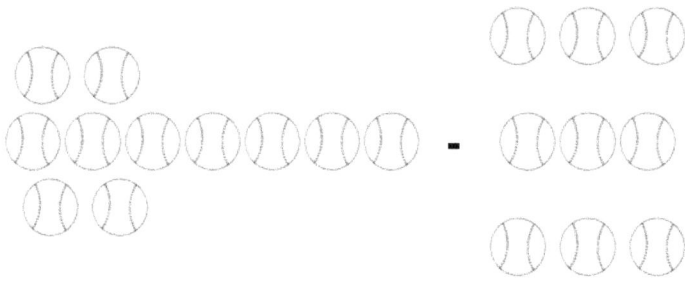

4   3   2

5. Subtract the crayons and then X the box that shows the right answer.

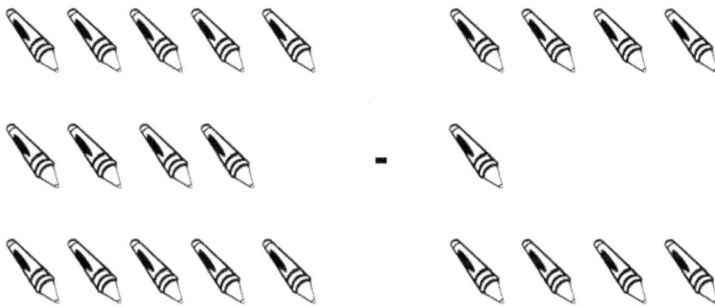

4   3   5

Unit 3 - Subtraction Using a Number Line

Instructor Note:

Your student should gain the ability to use a number line in place of concrete objects when solving subtraction problems.

Student Instructions:

Number lines can help you subtract. The number line will have a beginning number and an ending number.

This number line begins at the number 0 and ends at the number 24. That means that you can subtract any numbers between 0 and 24.

Use this number line to subtract 12 - 5

Place your pencil in the number 12.

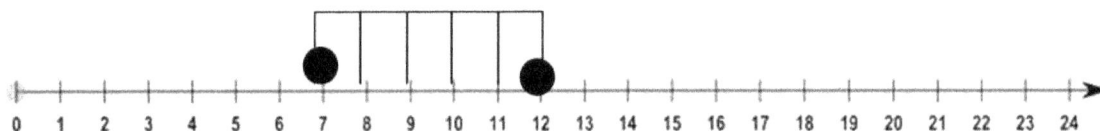

Jump your pencil backward 5 places landing on 12, 11, 10, 9, 8, and 7. The last number you land on is the answer.

$12 - 5 = 7$

Use the number line to solve each problem.

1.

    Use the number line below to solve 4 - 4 = _____

2.

Use the number line below to solve 18 - 3 = _____

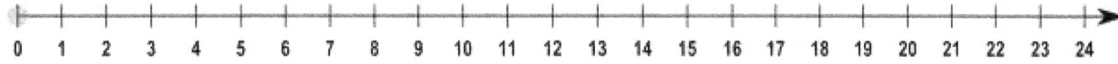

3.

Use the number line below to solve 10 - 5 = _____

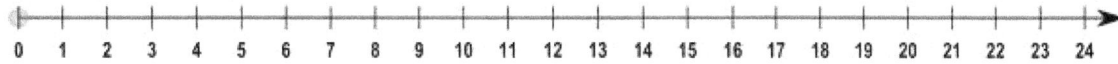

4.

Use the number line below to solve 6 - 6 = _____

5.

Use the number line below to solve 20 - 3 = _____

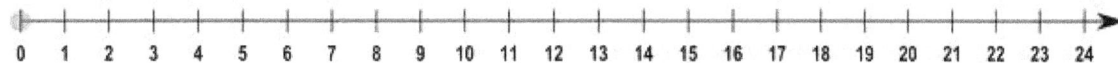

Name: _____

Use the number line to solve each problem.

1.

Use the number line below to solve 16 - 9 = _____

2.

Use the number line below to solve 8 - 8 = _____

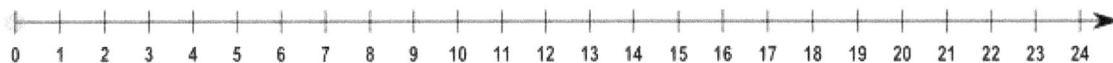

3.

Use the number line below to solve 7 - 5 = _____

4.

Use the number line below to solve 6 - 3 = _____

Name: _____

Use the number line to solve each problem.

1.

   Use the number line below to solve 19 - 5 = _____

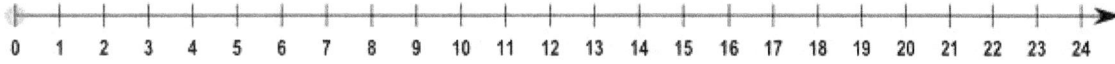

2.

   Use the number line below to solve 17 - 4 = _____

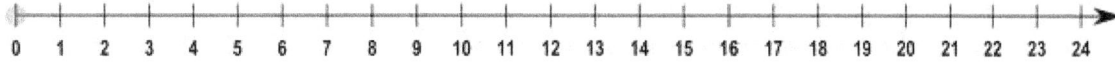

3.

   Use the number line below to solve 24 - 3 = _____

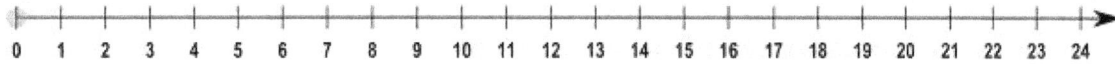

4.

   Use the number line below to solve 6 - 2 = _____

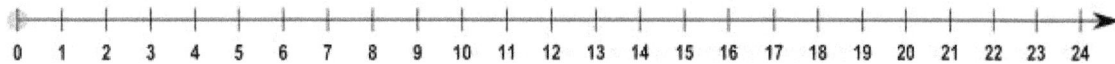

Name: _____

Use the number line to solve each problem.

1.

Use the number line below to solve 14 - 2 = _____

2.

Use the number line below to solve 7 - 3 = _____

3.

Use the number line below to solve 11 - 4 = _____

4.

Use the number line below to solve 5 - 5 = _____

Name: _____

Use the number line to solve each problem.

1.

   Use the number line below to solve 12 - 6 = _____

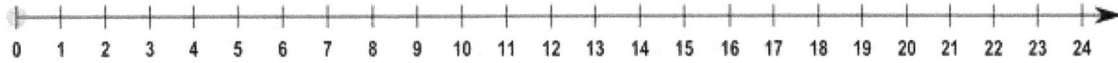

2.

   Use the number line below to solve 13 - 7 = _____

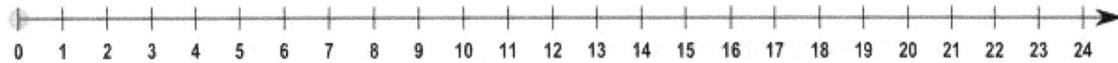

3.

   Use the number line below to solve 16 - 8 = _____

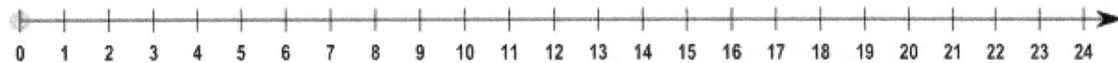

4.

   Use the number line below to solve 14 - 9 = _____

Name: _____

Use the number line to solve each problem.

1.

   Use the number line below to solve 24 - 8 = _____

2.

   Use the number line below to solve 19 - 9 = _____

3.

   Use the number line below to solve 12 - 7 = _____

4.

   Use the number line below to solve 18 - 6 = _____

Name: _____

Use the number line to solve each problem.

1.

   Use the number line below to solve 22 - 5 = _____

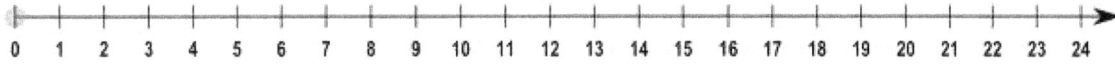

2.

   Use the number line below to solve 14 - 4 = _____

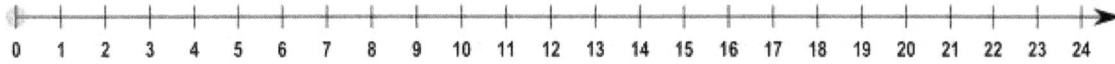

3.

   Use the number line below to solve 21 - 3 = _____

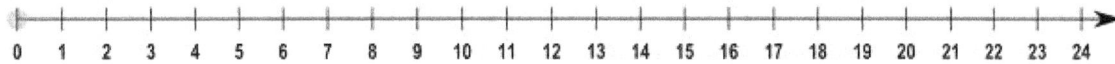

4.

   Use the number line below to solve 15 - 2 = _____

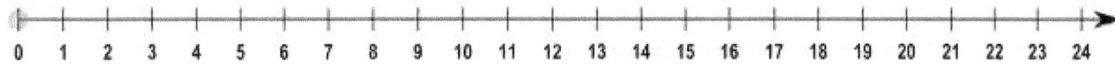

Name: _____

Use the number line to solve each problem.

1.

Use the number line below to solve 9 - 1 = _____

2.

Use the number line below to solve 18 - 12 = _____

3.

Use the number line below to solve 21 - 14 = _____

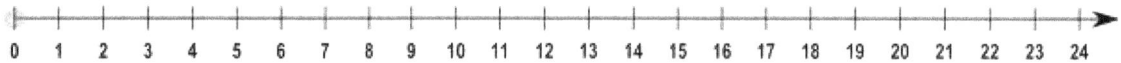

4.

Use the number line below to solve 16 - 13 = _____

Name: _____

Use the number line to solve each problem.

1.

Use the number line below to solve 19 - 6 = _____

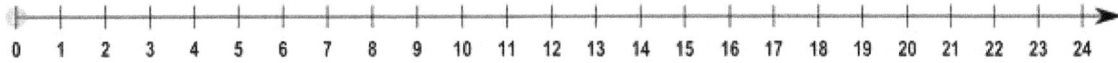

2.

Use the number line below to solve 17 - 17 = _____

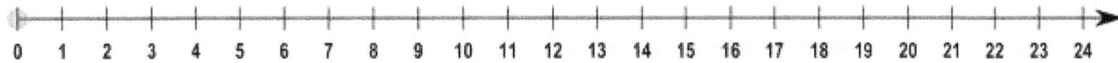

3.

Use the number line below to solve 22 - 8 = _____

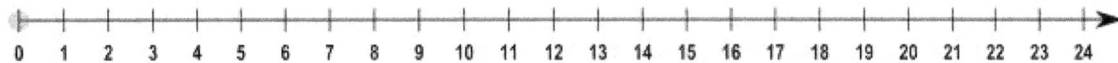

4.

Use the number line below to solve 15 - 9 = _____

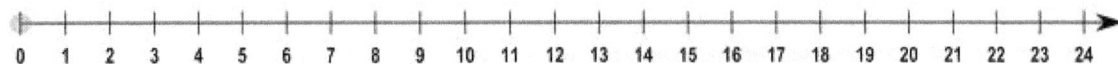

Name: _____

Use the number line to solve each problem.

1.

Use the number line below to solve 18 - 9 = _____

2.

Use the number line below to solve 23 - 8 = _____

3.

Use the number line below to solve 16 - 7 = _____

4.

Use the number line below to solve 24 - 6 = _____

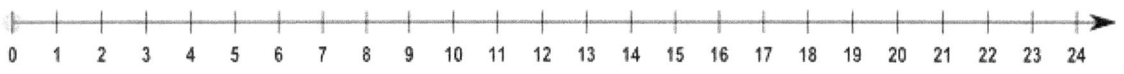

Unit 4 – Fact Families

Instructor Note:

Your student should gain an understanding of the inverse relationship between addition and subtraction using fact families.

Student Instructions:

Addition is the opposite operation of subtraction. They are related to one another.

1 + 2 = 3	2 + 1 = 3
3 – 2 = 1	3 – 1 = 2

This is known as a fact family. You have memorized your addition facts. If you take those same facts and reverse them, it makes memorizing your subtraction facts easier.

2 + 3 = 5	3 + 2 = 5		2 + 4 = 6	4 + 2 = 6		2 + 5 = 7	5 + 2 = 7
5 - 3 = 2	5 - 2 = 3		6 - 4 = 2	6 - 2 = 4		7 - 5 = 2	7 - 2 = 5

Complete the fact family charts.

2 + 6 = 8	2 + 7 = 9	2 + 8 = 10

2 + 9 = 11	3 + 1 = 4	3 + 2 = 5

3 + 3 = 6	3 + 4 = 7	3 + 5 = 8
3 + 6 = 9	3 + 7 = 10	3 + 8 = 11
3 + 9 = 12	4 + 1 = 5	4 + 2 = 6
4 + 3 = 7	4 + 4 = 8	4 + 5 = 9
4 + 6 = 10	4 + 7 = 11	4 + 8 = 12

4 + 9 = 13	5 + 1 = 6	5 + 2 = 7
5 + 3 = 8	5 + 4 = 9	5 + 5 = 10
5 + 6 = 11	5 + 7 = 12	5 + 8 = 13
5 + 9 = 14	6 + 1 = 7	6 + 2 = 8
6 + 3 = 9	6 + 4 = 10	6 + 5 = 11

6 + 6 = 12	6 + 7 = 13	6 + 8 = 14
6 + 9 = 15	7 + 1 = 8	7 + 2 = 9
7 + 3 = 10	7 + 4 = 11	7 + 5 = 12
7 + 6 = 13	7 + 7 = 14	7 + 8 = 15
7 + 9 = 16	8 + 1 = 9	8 + 2 = 10

8 + 3 = 11	8 + 4 = 12	8 + 5 = 13
8 + 6 = 14	8 + 7 = 15	8 + 8 = 16
8 + 9 = 17	9 + 1 = 10	9 + 2 = 11
9 + 3 = 12	9 + 4 = 13	9 + 5 = 14
9 + 6 = 15	9 + 7 = 16	9 + 8 = 17

Unit 5 – Subtraction Facts

Instructor Note:

Your student should memorize the basic subtraction facts so that they can subtract from memory instead of using tools. The subtraction fact chart and flash cards are excellent tools to assist you in helping your student master basic subtraction facts. You should also copy the pages from the unit transitioning to mental math and have your student complete these problems until they have memorized each subtraction fact.

Student Instructions:

Subtracting numbers is much easier if you memorize some subtraction facts. Knowing the answers to basic facts will help you to solve subtraction problems much more quickly.

Cut out the flash cards and practice your subtraction facts.

10 - 0	10 - 1	10 - 2
10 - 3	10 - 4	10 - 5
10 - 6	10 - 7	10 - 8
10 - 9	10 - 10	9 - 1

9 - 2	9 - 3	9 - 4
9 - 5	9 - 6	9 - 7
9 - 8	9 - 9	8 - 0
8 - 1	8 - 2	8 - 3
8 - 4	8 - 5	8 - 6
8 - 7	8 - 8	7 - 0

7 - 1	7 - 2	7 - 3
7 - 4	7 - 5	7 - 6
7 - 7	6 - 0	6 - 1
6 - 2	6 - 3	6 - 4
6 - 5	6 - 6	5 - 0
5 - 1	5 - 2	5 - 3
5 - 4	5 - 5	4 - 0

4 - 1	4 - 2	4 - 3
4 - 4	3 - 0	3 - 1
3 - 02	3 - 3	2 - 0
2 - 1	2 - 2	1 - 0
1 - 1		

Subtraction Facts

10-0=10   10-1=-   10-2=8   10-3=7   10-4=6   10-5=5   10-6=7   10-7=3   10-8=2   10-9=1

9-0=9   9-1=8   9-2=7   9-3=6   9-4=5   9-5=4   9-6=3   9-7=2   9-8=1   9-9=0

8-0=8   8-1=7   8-2=6   8-3=5   8-4=4   8-5=3   8-6=2   8-7=1   8-8=0

7-0=7   7-1=6   7-2=5   7-3=4   7-4=3   7-5=2   7-6=1   7-7=0

6-0=6   6-1=5   6-2=4   6-3=3   6-4=2   6-5=1   6-6=0

5-0=5   5-1=4   5-2=3   5-3=2   5-4=1   5-5=0

4-0=4   4-1=3   4-2=2   4-3=1   4-4=0

3-0=3   3-1=2   3-2=1   3-3=0

2-0=2   2-1=1   2-2=0

Subtraction Facts

Try It – Complete the subtraction facts chart.

10 - 0	10 - 1	10 - 2	10 - 3	10 - 4	10 - 5	10 - 6	10 - 7	10 - 8	10 - 9
10 - 0	9 -1	9 -2	9 -3	9 -4	9 -5	9 -6	9 -7	9 -8	9 -9
8 -0	8 -1	8 -2	8 -3	8 -4	8 -5	8 -6	8 -7	8 -8	7 - 0
7 -1	7 -2	7 -3	7 -4	7 -5	7 -6	7 -7	6 -0	6 -1	6 -2
6 -3	6 -4	6 -5	6 -6	5 -0	5 -1	5 -2	5 -3	5 -4	5 -5
4 -0	4 -1	4 -2	4 -3	4 -4	3 -0	3 -1	3 -2	3 -3	2 -2

Unit 6 – Subtraction – Transitioning to Mental Math

Instructor Note:

Your student should begin to transition concrete object modeling subtraction skills to number only subtraction skills.   You can accomplish the transition by providing subtraction problems with and then without visual aides.

Your student should move through the stages of completing subtraction with concrete objects, worksheet based concrete object aids and then with no visual aids whatsoever.

Student Instructions:

When you subtract one group of objects from another group of objects, what is left is a smaller group.

You can subtract numbers in your head the same way that you subtract pictures on the page.

Example: 5 - 3 = 2

Start at the number 5.

Hold the number 5 in your head.

Count backwards 3 numbers aloud.

4, 3, 2

You end up at 2 so 5 - 3 = 2.

Name: _____

Subtract the numbers and then pick the box with the correct answer to each problem.

1.
$$\begin{array}{r} 9 \\ -0 \\ \hline \end{array}$$

| 10 | 0 | 9 |

2.
$$\begin{array}{r} 5 \\ -0 \\ \hline \end{array}$$

| 0 | 4 | 5 |

3.
$$\begin{array}{r} 7 \\ -0 \\ \hline \end{array}$$

| 6 | 0 | 7 |

4.
$$\begin{array}{r} 2 \\ -0 \\ \hline \end{array}$$

| 2 | 3 | 0 |

5.
$$\begin{array}{r} 4 \\ -0 \\ \hline \end{array}$$

| 4 | 5 | 0 |

6.
$$\begin{array}{r} 3 \\ -0 \\ \hline \end{array}$$

| 4 | 3 | 0 |

7.
$$\begin{array}{r} 6 \\ -0 \\ \hline \end{array}$$

| 5 | 6 | 0 |

8.
$$\begin{array}{r} 8 \\ -0 \\ \hline \end{array}$$

| 9 | 8 | 0 |

Name: _____

Subtract the numbers and then pick the box with the correct answer to each problem.

1.　9
　　- 1

| 11 | 8 | 10 |

2.　5
　　- 1

| 3 | 4 | 6 |

3.　7
　　- 1

| 7 | 6 | 8 |

4.　2
　　- 1

| 3 | 4 | 1 |

5.　4
　　- 1

| 5 | 6 | 3 |

6.　3
　　- 1

| 2 | 4 | 3 |

7.　6
　　- 1

| 5 | 6 | 7 |

8.　8
　　- 1

| 9 | 7 | 10 |

Name: _____

Subtract the numbers and then pick the box with the correct answer to each problem.

1. 9
   -2

    [7] [9] [11]

2. 5
   -2

    [4] [5] [3]

3. 7
   -2

    [5] [6] [9]

4. 2
   -2

    [3] [4] [0]

5. 4
   -2

    [3] [6] [2]

6. 3
   -2

    [5] [4] [1]

7. 6
   -2

    [8] [6] [4]

8. 8
   -2

    [10] [9] [6]

Name: _____

Subtract the numbers and then pick the box with the correct answer to each problem.

1.   9
    -3

    [6]   [5]   [12]

2.   5
    -3

    [2]   [3]   [8]

3.   7
    -3

    [4]   [10]   [5]

4.   10
    -3

    [7]   [4]   [5]

5.   4
    -3

    [1]   [0]   [7]

6.   3
    -3

    [0]   [1]   [6]

7.   6
    -3

    [3]   [9]   [4]

8.   8
    -3

    [5]   [6]   [11]

Name: _____

Subtract the numbers and then pick the box with the correct answer to each problem.

1. 9
   − 4

   ⬚ 5   ⬚ 6   ⬚ 13

2. 5
   − 4

   ⬚ 1   ⬚ 2   ⬚ 9

3. 7
   − 4

   ⬚ 3   ⬚ 11   ⬚ 5

4. 12
   − 4

   ⬚ 7   ⬚ 8   ⬚ 6

5. 4
   − 4

   ⬚ 0   ⬚ 1   ⬚ 8

6. 10
   − 4

   ⬚ 8   ⬚ 6   ⬚ 7

7. 6
   − 4

   ⬚ 2   ⬚ 3   ⬚ 5

8. 8
   − 4

   ⬚ 4   ⬚ 3   ⬚ 2

Name: _____

Solve each problem.

1.   9
    -5

2.   14
     -5

3.   7
    -5

4.   12
     -5

5.   11
    -5

6.   10
    -5

7.   6
    -5

8.   8
    -5

9.   13
    -5

10.   5
     -5

Name: _____

Solve each problem.

1.
$$\begin{array}{r} 9 \\ -6 \\ \hline \end{array}$$

2.
$$\begin{array}{r} 11 \\ -6 \\ \hline \end{array}$$

3.
$$\begin{array}{r} 14 \\ -6 \\ \hline \end{array}$$

4.
$$\begin{array}{r} 12 \\ -6 \\ \hline \end{array}$$

5.
$$\begin{array}{r} 10 \\ -6 \\ \hline \end{array}$$

6.
$$\begin{array}{r} 7 \\ +6 \\ \hline \end{array}$$

7.
$$\begin{array}{r} 6 \\ -6 \\ \hline \end{array}$$

8.
$$\begin{array}{r} 8 \\ -6 \\ \hline \end{array}$$

9.
$$\begin{array}{r} 13 \\ -6 \\ \hline \end{array}$$

10.
$$\begin{array}{r} 7 \\ -6 \\ \hline \end{array}$$

Name: _____

Solve each problem.

1. 14
   -7

2. 10
   -7

3.   7
   -7

4. 12
   -7

5. 11
   -7

6.   8
   -7

7. 13
   -7

8.   9
   -7

9. 16
   -7

10. 17
   -7

Name: _____

Solve each problem.

1. 9
   -8

2. 10
   -8

3. 12
   -8

4. 8
   -8

5. 11
   -8

6. 9
   -9

7. 11
   -9

8. 10
   -9

9. 11
   -10

10. 12
   -10

Unit 7 – Number Lines and Bigger Numbers

Instructor Note:

Your student should gain the ability to use a number line in place of concrete objects to help solve bigger number subtraction problems.

Student Instructions:

Number lines can help you subtract bigger numbers too. The number line will have a beginning number and an ending number.

This number line begins at the number 0 and ends at the number 24. That means that you can subtract any two numbers between 0 and 24. You can make a number line that starts and ends at any number.

Use this number line to subtract 23 – 11.

Place your pencil in the number 23.

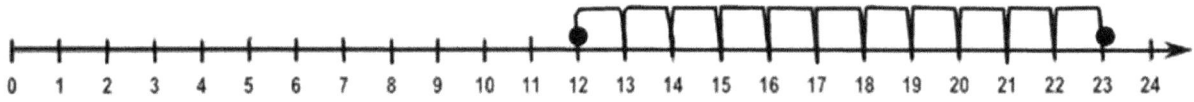

Jump your pencil backward 11 places landing on 22, 21, 20, 19, 18, 17, 16, 15, 14, 13, and 12. The last number you land on is the answer.

23 - 11 = 12

Use the number line to solve each problem.

1.

Use the number line below to solve 24 - 9 = _____

66

2.

Use the number line below to solve 15 - 4 = _____

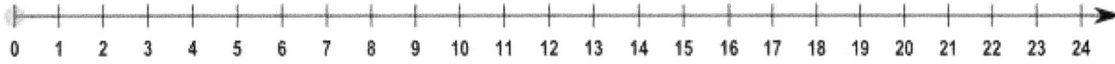

3.

Use the number line below to solve 19 - 11 = _____

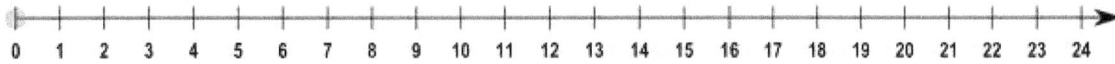

4.

Use the number line below to solve 21 - 12 = _____

5.

Use the number line below to solve 23 - 8 = _____

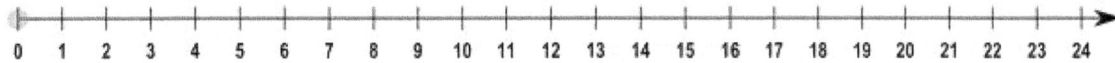

Name: _____

Use the number line to solve each problem.

1.

    Use the number line below to solve 20 - 9 = _____

```
├──┼──┼──┼──┼──┼──┼──┼──┼──┼──┼──┼──┼──┼──┼──┼──┼──┼──┼──┼──┼──┼──┼──┼──►
0   1   2   3   4   5   6   7   8   9  10  11  12  13  14  15  16  17  18  19  20  21  22  23  24
```

2.

    Use the number line below to solve 16 - 8 = _____

```
├──┼──┼──┼──┼──┼──┼──┼──┼──┼──┼──┼──┼──┼──┼──┼──┼──┼──┼──┼──┼──┼──┼──┼──►
0   1   2   3   4   5   6   7   8   9  10  11  12  13  14  15  16  17  18  19  20  21  22  23  24
```

3.

    Use the number line below to solve 12 - 5 = _____

```
├──┼──┼──┼──┼──┼──┼──┼──┼──┼──┼──┼──┼──┼──┼──┼──┼──┼──┼──┼──┼──┼──┼──┼──►
0   1   2   3   4   5   6   7   8   9  10  11  12  13  14  15  16  17  18  19  20  21  22  23  24
```

4.

    Use the number line below to solve 18 - 14 = _____

```
├──┼──┼──┼──┼──┼──┼──┼──┼──┼──┼──┼──┼──┼──┼──┼──┼──┼──┼──┼──┼──┼──┼──┼──►
0   1   2   3   4   5   6   7   8   9  10  11  12  13  14  15  16  17  18  19  20  21  22  23  24
```

Name: _____

Use the number line to solve each problem.

1.

Use the number line below to solve 19 - 13 = _____

2.

Use the number line below to solve 17 - 4 = _____

3.

Use the number line below to solve 16 - 12 = _____

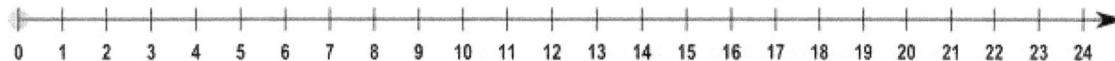

4.

Use the number line below to solve 22 - 2 = _____

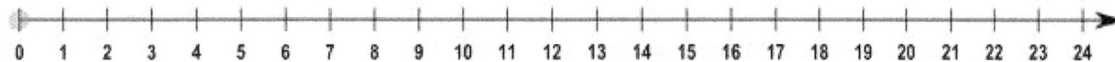

Name: _____

Use the number line to solve each problem.

1.

Use the number line below to solve 18 - 4 = _____

2.

Use the number line below to solve 17 - 11 = _____

3.

Use the number line below to solve 19 - 5 = _____

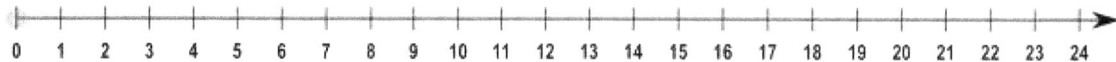

4.

Use the number line below to solve 15 - 15 = _____

Name: _____

Use the number line to solve each problem.

1.

Use the number line below to solve 20 - 12 = _____

2.

Use the number line below to solve 21 - 3 = _____

3.

Use the number line below to solve 18 - 11 = _____

4.

Use the number line below to solve 23 - 20 = _____

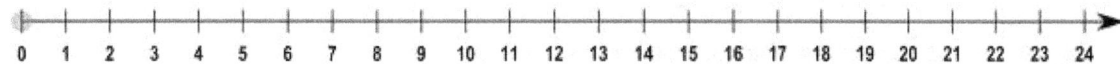

Unit 8 – Subtraction Using a 100's Chart

Instructor Note:

Your student should transition to higher number subtraction with the aid of a hundred chart.

Student Instructions:

A 100's chart can help you to subtract larger numbers. The 100's chart begins at the number 1 and ends at the number 100. That means that you can use the hundred chart to help you subtract any numbers between 1 and 100.

Use the 100's chart to subtract 62 - 12

1	2	3	4	5	6	7	8	9	10
11	12	13	14	15	16	17	18	19	20
21	22	23	24	25	26	27	28	29	30
31	32	33	34	35	36	37	38	39	40
41	42	43	44	45	46	47	48	49	50
51	52	53	54	55	56	57	58	59	60
61	62	63	64	65	66	67	68	69	70
71	72	73	74	75	76	77	78	79	80
81	82	83	84	85	86	87	88	89	90
91	92	93	94	95	96	97	98	99	100

Place your pencil on the number 62.

Jump your pencil backward 12 places landing on 61, 60, 59, 58, 57, 56, 55, 54, 53, 52, 51 and 52. The last number you land on is the answer.

62 - 12 = 50

1	2	3	4	5	6	7	8	9	10
11	12	13	14	15	16	17	18	19	20
21	22	23	24	25	26	27	28	29	30
31	32	33	34	35	36	37	38	39	40
41	42	43	44	45	46	47	48	49	50
51	52	53	54	55	56	57	58	59	60
61	62	63	64	65	66	67	68	69	70
71	72	73	74	75	76	77	78	79	80
81	82	83	84	85	86	87	88	89	90
91	92	93	94	95	96	97	98	99	100

Cut out the 100's chart and use it to solve each problem.

Name: _____

Use the 100 chart to help you solve each problem.

1. 70
    $\underline{-6}$

2. 55
    $\underline{-2}$

3. 18
    $\underline{-5}$

4. 22
    $\underline{-3}$

5. 28
    $\underline{-9}$

6. 38
    $\underline{-4}$

7. 31
    $\underline{-7}$

8. 61
    $\underline{-8}$

9. 25
    $\underline{-9}$

10. 42
    $\underline{-8}$

Name: _____

Use the 100 chart to help you solve each problem.

1. $\begin{array}{r} 61 \\ -5 \\ \hline \end{array}$ 2. $\begin{array}{r} 46 \\ -2 \\ \hline \end{array}$

3. $\begin{array}{r} 29 \\ -5 \\ \hline \end{array}$ 4. $\begin{array}{r} 33 \\ -3 \\ \hline \end{array}$

5. $\begin{array}{r} 17 \\ -9 \\ \hline \end{array}$ 6. $\begin{array}{r} 56 \\ -4 \\ \hline \end{array}$

7. $\begin{array}{r} 62 \\ -7 \\ \hline \end{array}$ 8. $\begin{array}{r} 74 \\ -8 \\ \hline \end{array}$

9. $\begin{array}{r} 95 \\ -3 \\ \hline \end{array}$ 10. $\begin{array}{r} 87 \\ -8 \\ \hline \end{array}$

Name: _____

Use the 100 chart to help you solve each problem.

1. 59
   -21

2. 32
   -30

3. 87
   -11

4. 64
   -19

5. 21
   -16

6. 19
   -11

7. 46
   -12

8. 78
   -14

9. 90
   -8

10. 53
    -18

Name: _____

Use the 100 chart to help you solve each problem.

1. 70
   -25

2. 55
   - 32

3. 27
   -18

4. 34
   - 22

5. 48
   -29

6. 38
   -26

7. 31
   -23

8. 61
   - 18

9. 65
   -39

10. 42
   - 38

Name: _____

Use the 100 chart to help you solve each problem.

1.   67
     -25

2.   42
     - 32

3.   38
     -27

4.   66
     - 34

5.   54
     -29

6.   71
     -26

7.   80
     -23

8.   64
     - 18

9.   25
     -25

10.   51
     - 38

Unit 9 – 2 Digit Subtraction

Some problems will have more than 1 digit on each line. You can do all of these problems using the math facts you have learned!

$$\begin{array}{r} 25 \\ -\,12 \\ \hline \end{array}$$

You will solve these problems starting at the ones place and moving to the left.

$$\begin{array}{r} 25 \\ -\,12 \\ \hline 3 \end{array} \qquad\qquad \begin{array}{r} 25 \\ -\,12 \\ \hline 13 \end{array}$$

Solve each problem using your math facts.

1.
$$\begin{array}{r} 56 \\ -\,11 \\ \hline \end{array}$$

2.
$$\begin{array}{r} 48 \\ -\,16 \\ \hline \end{array}$$

3.
$$\begin{array}{r} 36 \\ -\,12 \\ \hline \end{array}$$

4.
$$\begin{array}{r} 45 \\ -\,32 \\ \hline \end{array}$$

5.
$$\begin{array}{r} 78 \\ -\,14 \\ \hline \end{array}$$

6.
$$\begin{array}{r} 87 \\ -\,66 \\ \hline \end{array}$$

7.
$$\begin{array}{r} 65 \\ -\,22 \\ \hline \end{array}$$

8.
$$\begin{array}{r} 98 \\ -\,27 \\ \hline \end{array}$$

9.
$$\begin{array}{r} 29 \\ -\,15 \\ \hline \end{array}$$

Name: _____

Solve each problem using your math facts.

1.   56
       − 11

2.   47
       − 16

3.   36
       − 12

4.   76
       − 32

5.   45
       − 14

6.   89
       − 66

7.   65
       − 22

8.   78
       − 27

9.   46
       − 15

10.   65
       − 22

11.   39
       − 27

12.   85
       − 13

Name: _____

Solve each problem using your math facts.

1.　　45
　　 − 21

2.　　79
　　 − 40

3.　　85
　　 − 61

4.　　36
　　 − 21

5.　　94
　　 − 53

6.　　67
　　 − 55

7.　　54
　　 − 11

8.　　67
　　 − 16

9.　　76
　　 − 24

10.　　96
　　 − 23

11.　　59
　　 − 40

12.　　85
　　 − 12

Name: _____

Solve each problem using your math facts.

1.　　　34
　　　　－ 32

2.　　　67
　　　　－ 22

3.　　　56
　　　　－ 44

4.　　　87
　　　　－ 32

5.　　　45
　　　　－ 42

6.　　　66
　　　　－ 32

7.　　　65
　　　　－ 22

8.　　　76
　　　　－ 21

9.　　　83
　　　　－ 13

10.　　　57
　　　　－ 32

11.　　　79
　　　　－ 20

12.　　　87
　　　　－ 11

Name: _____

Solve each problem using your math facts.

1.   75
        - 33

2.   58
        - 21

3.   67
        - 42

4.   48
        -11

5.   96
        - 31

6.   85
        - 55

7.   26
        - 11

8.   87
        - 66

9.   15
        - 14

10.   68
        - 21

11.   39
        - 18

12.   19
        - 16

Unit 9 – 3 Digit Subtraction

Some problems will have more than 1 digit on each line. You can do all of these problems using the math facts you have learned!

$$\begin{array}{r} 537 \\ -\ 321 \\ \hline \end{array}$$

You will solve these problems starting at the ones place and moving to the left.

$$\begin{array}{r} 537 \\ -\ 321 \\ \hline 6 \end{array} \qquad \begin{array}{r} 537 \\ -\ 321 \\ \hline 16 \end{array} \qquad \begin{array}{r} 537 \\ -\ 321 \\ \hline 216 \end{array}$$

Solve each problem using your math facts.

1. $\begin{array}{r} 354 \\ -\ 102 \\ \hline \end{array}$ 2. $\begin{array}{r} 236 \\ -\ 211 \\ \hline \end{array}$ 3. $\begin{array}{r} 356 \\ -\ 142 \\ \hline \end{array}$

4. $\begin{array}{r} 536 \\ -\ 422 \\ \hline \end{array}$ 5. $\begin{array}{r} 445 \\ -\ 214 \\ \hline \end{array}$ 6. $\begin{array}{r} 669 \\ -\ 336 \\ \hline \end{array}$

7. $\begin{array}{r} 465 \\ -\ 122 \\ \hline \end{array}$ 8. $\begin{array}{r} 779 \\ -\ 227 \\ \hline \end{array}$ 9. $\begin{array}{r} 885 \\ -\ 113 \\ \hline \end{array}$

Name: _____

Solve each problem using your math facts.

1.   536
         - 131

2.   466
         - 120

3.   356
         - 142

4.   756
         - 232

5.   845
         - 814

6.   696
         - 302

7.   565
         - 322

8.   797
         - 201

9.   485
         - 312

10.   685
         - 212

11.   577
         - 521

12.   855
         - 143

Name: _____

Solve each problem using your math facts.

1.	345 - 221	2.	649 - 230	3.	695 - 201

4.	826 - 101	5.	574 - 323	6.	758 - 125

7.	654 - 311	8.	676 - 126	9.	754 - 242

10.	576 - 423	11.	659 - 340	12.	486 - 412

Name: _____

Solve each problem using your math facts.

1. 374
    - 332

2. 697
    - 202

3. 456
    - 342

4. 367
    - 232

5. 495
    - 402

6. 766
    - 232

7. 765
    - 122

8. 755
    - 241

9. 683
    - 313

10. 557
    - 432

11. 879
    - 123

12. 867
    - 342

Name: _____

Solve each problem using your math facts.

1.	495 - 303	2.	758 - 231	3.	557 - 342

4.	478 - 321	5.	656 - 231	6.	445 - 435

7.	876 - 121	8.	586 - 312	9.	625 - 224

10.	568 - 321	11.	679 - 120	12.	756 - 232

Unit 10 – Lining Up Numbers for Subtraction

Instructor Note:

Your student should understand how to line up horizontal subtraction problems in a vertical format.

Student Instructions:

Sometimes a math problem will appear in a line.

38 - 17 =

Other times a math problem will appear in a stack.

```
    3   8
-   1   7
_____
```

You should always rearrange the problem so that it lines up in a stack. This will help you to solve larger number math problems.

Write the first number in the math problem on your paper.

```
    3   8
```

Write the second number underneath so that the ones place of each number lines up with the one above it.

```
    3   8
    1   7
```

Add the symbol that tells you this is a subtraction problem.

```
    3   8
-   1   7
```

Draw a line under the bottom number to show where the answer to the problem begins.

```
    3   8
-   1   7
_____
```

Now the problem is lined up so that you know which numbers to subtract.

You will subtract the numbers starting with the smallest place value and calculating one column at a time to the left to the largest place value.

You will begin by subtracting the ones column.

```
      3    8
 -    1    7
 _____
           1
```

Next, you will subtract the numbers in the tens column.

```
      3    8
 -    1    7
 _____
      2    1
```

If the numbers were larger, you would continue working right to left until you have completed the calculations for each column.

Name: _____

Line up each problem and then solve it.

1.
$14 - 5 =$ _____

2.
$15 - 4 =$ _____

3.
$19 - 8 =$ _____

4.
$17 - 7 =$ _____

5.
$16 - 5 =$ _____

Name: _____

Line up each problem and then solve it.

1.

   10 - 9 = _____

_____

2.

   16 - 12 = _____

_____

3.

   17 - 15 = _____

_____

4.

   17 - 12 = _____

_____

5.

   18 - 13 = _____

_____

Name: _____

Line up each problem and then solve it.

1.
   26 - 15 = _____

_____

2.
   75 - 24 = _____

_____

3.
   59 - 38 = _____

_____

4.
   38 - 27 = _____

_____

5.
   86 - 55 = _____

_____

Name: _____

Line up each problem and then solve it.

1.

635 - 24 = _____

_____

2.

275 - 24 = _____

_____

3.

458 - 41 = _____

_____

4.

757 - 32 = _____

_____

5.

485 - 13 = _____

_____

Name: _____

Line up each problem and then solve it.

1.
   549 - 210 = _____

_____

2.
   686 - 312 = _____

_____

3.
   865 - 122 = _____

_____

4.
   437 - 322 = _____

_____

5.
   738 - 221 = _____

_____

Name: _____

Line up each problem and then solve it.

1.

439 - 118 = _____

_____

2.

575 - 211 = _____

_____

3.

784 - 251 = _____

_____

4.

326 - 221 = _____

_____

5.

627 - 110 = _____

_____

Unit 11 – Subtracting Larger Numbers

Instructor Note:

Your student should transition from subtracting one and two digit numbers to subtracting larger numbers.

Student Instructions:

Subtracting larger numbers is the same as subtracting smaller numbers. You will always begin at the right with the unit column and move to the left subtracting the tens, hundreds, thousands, ten thousands, hundred thousands, millions, and ten millions.

Remember your place value chart so that you know what the value of each column is before you add.

61,361,986									
Ten Millions	Millions	,	Hundred Thousands	Ten Thousands	Thousands	,	Hundreds	Tens	Ones
Six Ten Millions	One Million	,	Three Hundred Thousands	Six Ten Thousands	One Thousand	,	Nine Hundreds	Eight Tens	Six Ones
60,000,000	1,000,000	,	300,000	60,000	1,000	,	900	80	3

Remember to add the commas in the correct location to help to make your answers easier to read. A comma goes after every 3 places moving from right to left.

Name: _____

1. 457
    − 25

2. 249
    − 30

3. 138
    − 21

4. 536
    − 32

5. 654
    − 23

6. 977
    − 12

7. 845
    − 23

8. 665
    − 35

9. 729
    − 26

10. 257
    − 34

Name: _____

1. 359
   - 36

2. 158
   - 41

3. 137
   - 32

4. 465
   - 43

5. 557
   - 34

6. 876
   - 23

7. 744
   - 34

8. 568
   - 26

9. 648
   - 37

10. 257
    - 45

Name: _____

1. 456
   −336

2. 947
   −741

3. 638
   −630

4. 554
   −443

5. 653
   −332

6. 775
   −223

7. 844
   −134

8. 963
   − 820

9. 729
   −200

10. 356
   − 245

Name: _____

1.   657
     -437

2.   748
     - 641

3.   139
     -130

4.   553
     -542

5.   756
     -235

6.   674
     -124

7.   543
     -323

8.   867
     - 624

9.   678
     -261

10.   452
     - 241

102

Name: _____

1. 7,457
   -   25

2. 4,249
   -   30

3. 5,138
   -   21

4. 1,576
   -   32

5. 9,654
   -   23

6. 3,977
   -   12

7. 2,845
   -   23

8. 8,664
   -   32

9. 6,726
   -   15

10. 7,255
    -   34

1. $$\begin{array}{r} 5{,}357 \\ -\ \ 36 \\ \hline \end{array}$$

2. $$\begin{array}{r} 9{,}178 \\ -\ \ 41 \\ \hline \end{array}$$

3. $$\begin{array}{r} 3{,}137 \\ -\ \ 32 \\ \hline \end{array}$$

4. $$\begin{array}{r} 6{,}465 \\ -\ \ 43 \\ \hline \end{array}$$

5. $$\begin{array}{r} 7{,}556 \\ -\ \ 34 \\ \hline \end{array}$$

6. $$\begin{array}{r} 4{,}876 \\ -\ \ 23 \\ \hline \end{array}$$

7. $$\begin{array}{r} 1{,}744 \\ -\ \ 34 \\ \hline \end{array}$$

8. $$\begin{array}{r} 3{,}569 \\ -\ \ 26 \\ \hline \end{array}$$

9. $$\begin{array}{r} 2{,}687 \\ -\ \ 31 \\ \hline \end{array}$$

10. $$\begin{array}{r} 8{,}258 \\ -\ \ 41 \\ \hline \end{array}$$

Name: _____

1. 4,669
   - 536

2. 7,957
   - 741

3. 8,738
   - 630

4. 1,564
   - 443

5. 3,655
   - 334

6. 5,775
   - 223

7. 9,844
   - 134

8. 2,863
   - 823

9. 6,729
   - 220

10. 4,358
    - 345

Name: _____

1. 2,759
   - 548

2. 8,968
   - 751

3. 5,999
   - 640

4. 7,313
   - 201

5. 1,645
   - 324

6. 9,564
   - 234

7. 3,567
   - 433

8. 6,756
   - 533

9. 9,548
   - 311

10. 4,657
   - 642

Name: _____

1. 4,556
  − 430

2. 7,747
  − 231

3. 8,638
  − 537

4. 1,544
  − 443

5. 3,654
  − 332

6. 5,775
  − 223

7. 9,844
  − 134

8. 2,866
  − 423

9. 6,769
  − 220

10. 4,655
  − 343

Name: _____

1. 6,545
   -3,142

2. 7,797
   - 2,201

3. 6,648
   -2,130

4. 8,455
   -1,423

5. 7,545
   -2,332

6. 5,654
   -3.233

7. 7,562
   -1,422

8. 8,366
   -1,131

9. 4,649
   -2,320

10. 5,556
    - 4,343

Name: _____

1. 39,556
   − 9,430

2. 55,747
   − 4,211

3. 14,638
   − 3,230

4. 86,544
   − 1,423

5. 63,654
   − 1,333

6. 29,775
   − 7,223

7. 58,844
   − 5,134

8. 47,866
   − 2,423

9. 78,769
   − 6,220

10. 24,655
    − 2,343

Name: _____

1.   72,548
     -12,231

2.   48,758
     -31,108

3.   55,649
     -23,020

4.   78,653
     -17,313

5.   89,654
     -10,332

6.   69,564
     -30,234

7.   56,563
     -33,433

8.   46,756
     - 43,233

9.   49,548
     -12,311

10.   76,657
     - 12,342

Name: _____

1.   169,556
    - 30,532

2.   455,747
    - 24,211

3.   774,638
    - 13,330

4.   386,544
    - 11,423

5.   563,654
    - 31,333

6.   269,775
    - 27,223

7.   655,844
    - 41,134

8.   857,866
    - 42,123

9.   276,769
    - 22,220

10.   535,655
    - 25,343

Name: _____

1. 572,548
 -402,231

2. 858,758
 -851,101

3. 755,649
 -113,020

4. 647,653
 -511,313

5. 789,645
 -206,324

6. 669,564
 -120,234

7. 488,567
 -366,433

8. 856,758
 -143,233

9. 579,548
 -310,311

10. 676,657
 -202,335

Name: _____

1. 5,869,556
   - 130,430

2. 7,555,747
   - 224,211

3. 3,774,638
   - 213,230

4. 1,686,554
   - 312,433

5. 4,563,655
   - 431,323

6. 8,569,775
   - 220,223

7. 2,655,844
   - 141,134

8. 6,857,864
   - 142,123

9. 4,376,769
   - 221,250

10. 1,534,655
    - 412,443

Unit 12 – Borrowing in Subtraction

Instructor Note:

You student will need to use their knowledge of units, tens, hundreds, and thousands to gain the ability to regroup (borrow) within subtraction problems.

Student Instructions:

You should always line subtraction problems up so that the place value of the numbers line up one on top of the other. 231 - 116 becomes

```
    2    3    1
-   1    1    6
_____
```

When subtracting numbers, you will always start with the numbers at the right. This is the unit number.

This problem is asking us to subtract 6 from 1. You cannot subtract 6 from 1 because 1 is a smaller number.

```
    1
-   6
_____
```

You will need to borrow from the tens in order to finish this problem.

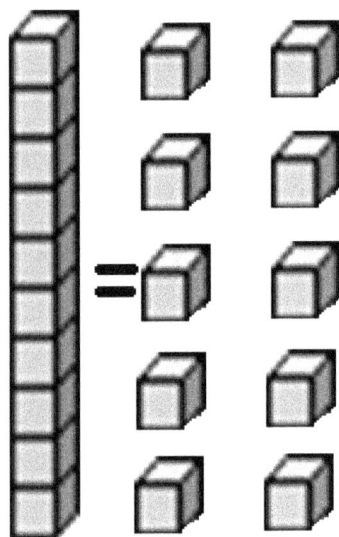

Remember, one 10 is equal to 10 units.

When you borrow in math, you are taking 1 ten from the tens column and breaking it back into units to add to the units column

```
          2    11
    2    X    1
  -  1    1    6
```

To remind ourselves that we borrowed, we write a little number above the column we borrowed from and above the column where we moved the borrowed digits.

After you have borrowed, the problem is solvable.

Hundreds	Tens	Ones
2	2	11
- 1	1	6
1	1	5

Understanding borrowing is easier if you remember your base 10 blocks. When you worked with addition, you regrouped the numbers into the largest block that they could make. When you borrow for subtraction, you break the blocks down into smaller units in order to make subtracting easier.

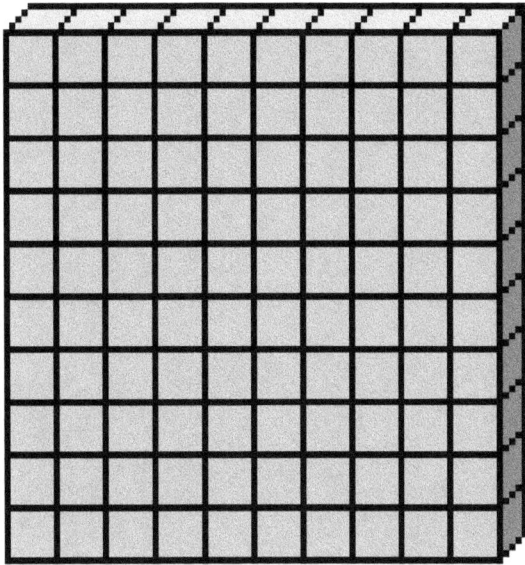
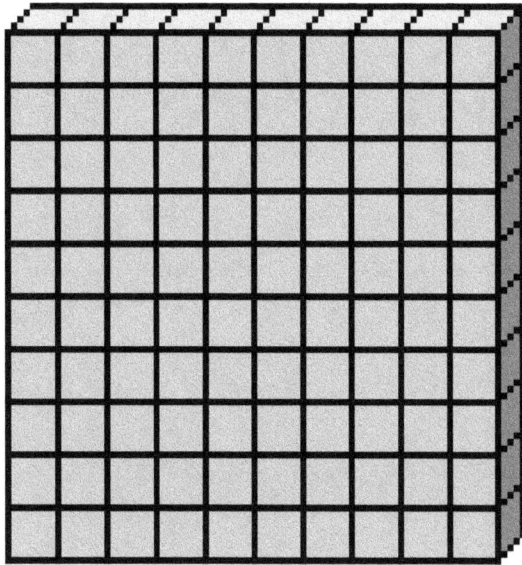
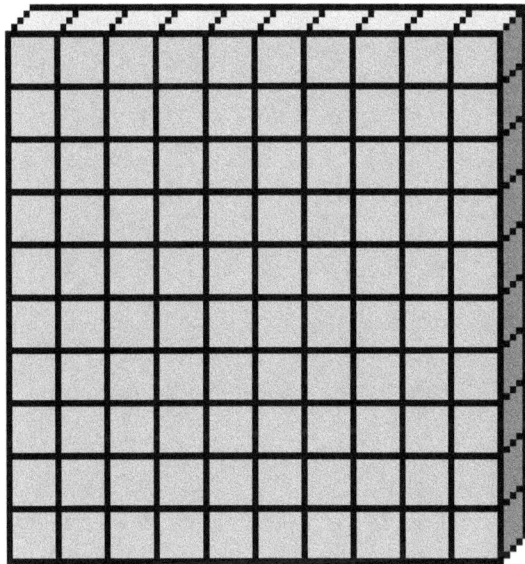

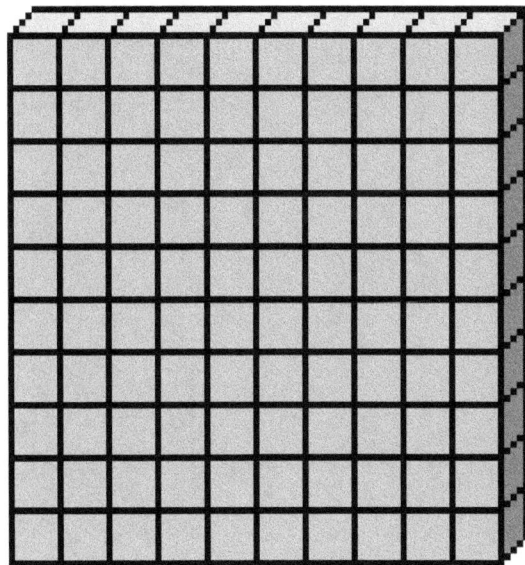
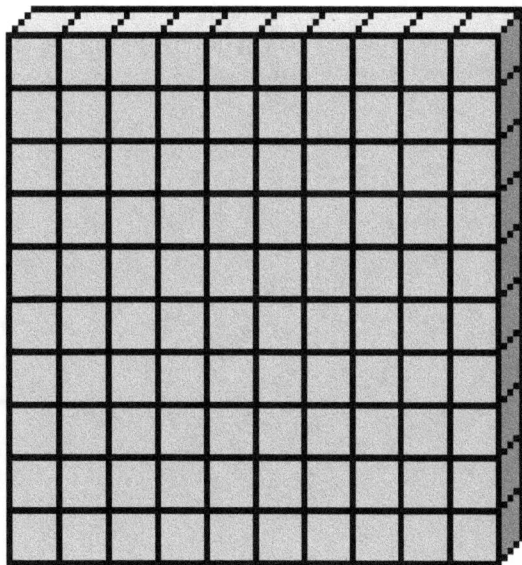

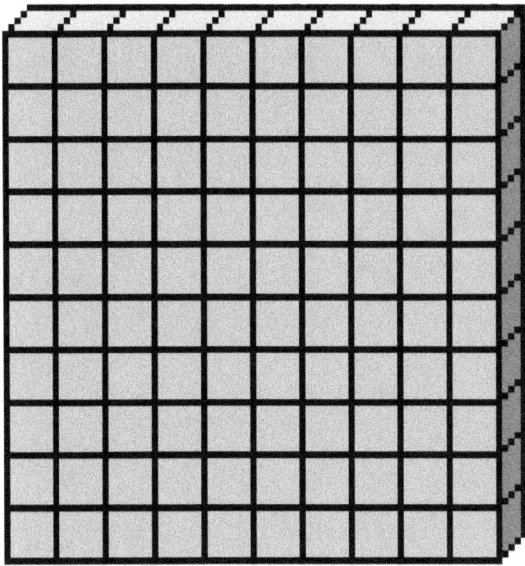
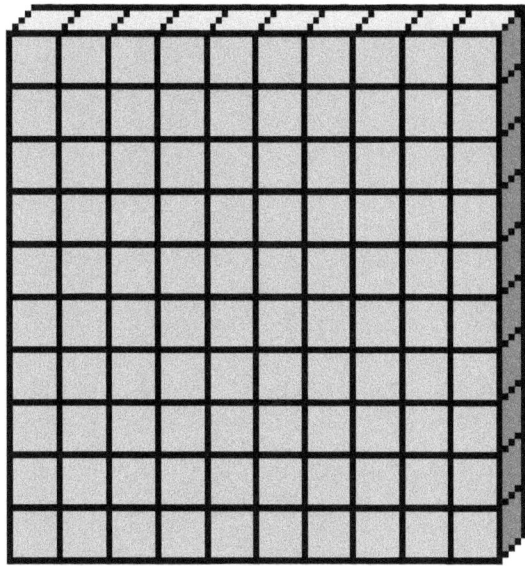
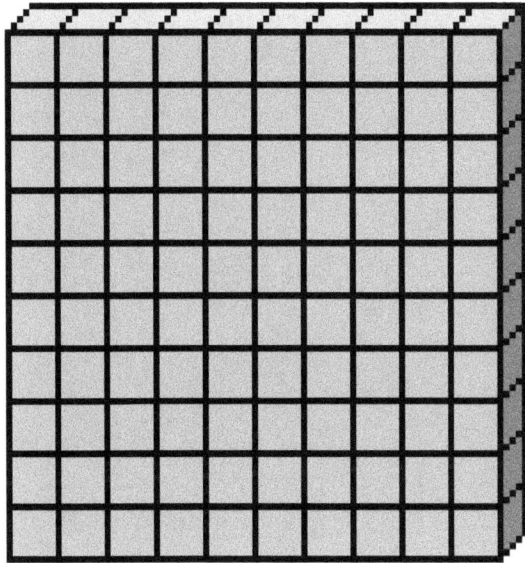
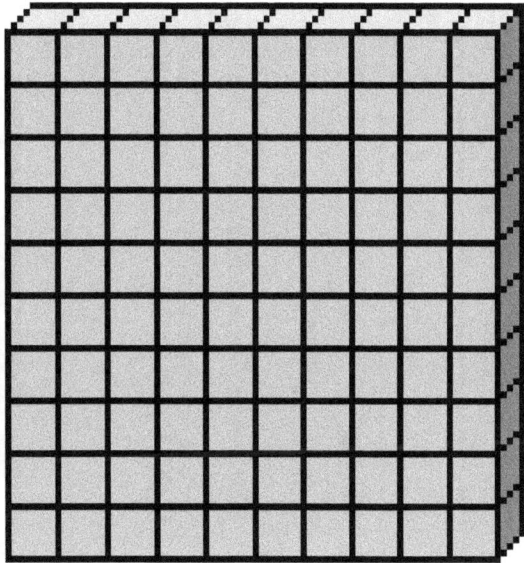
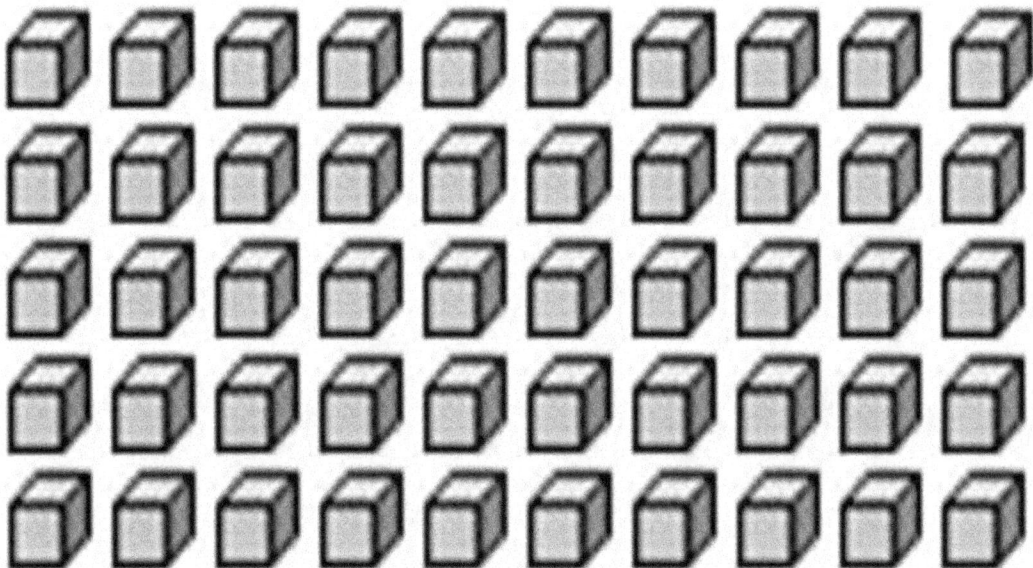

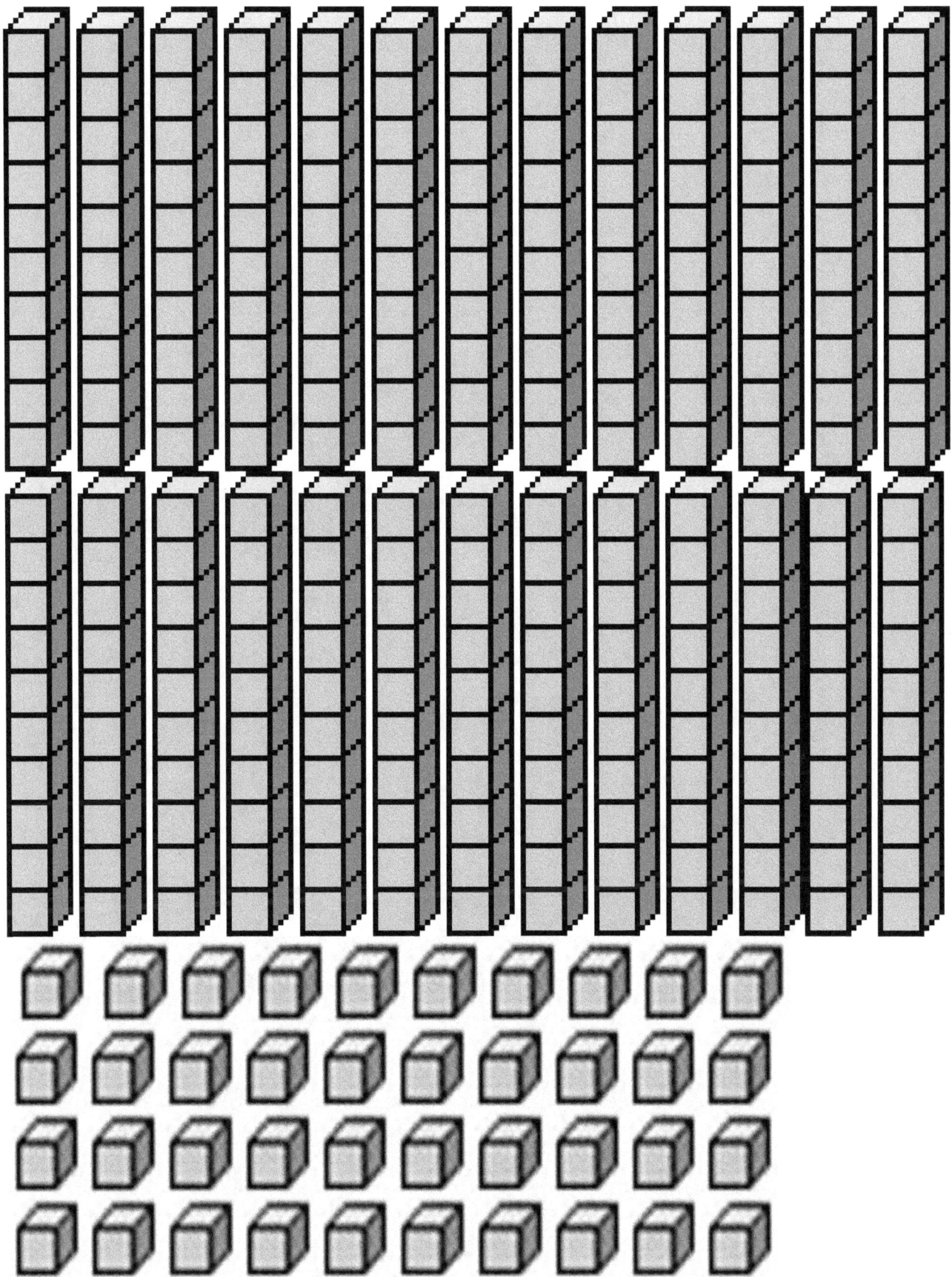

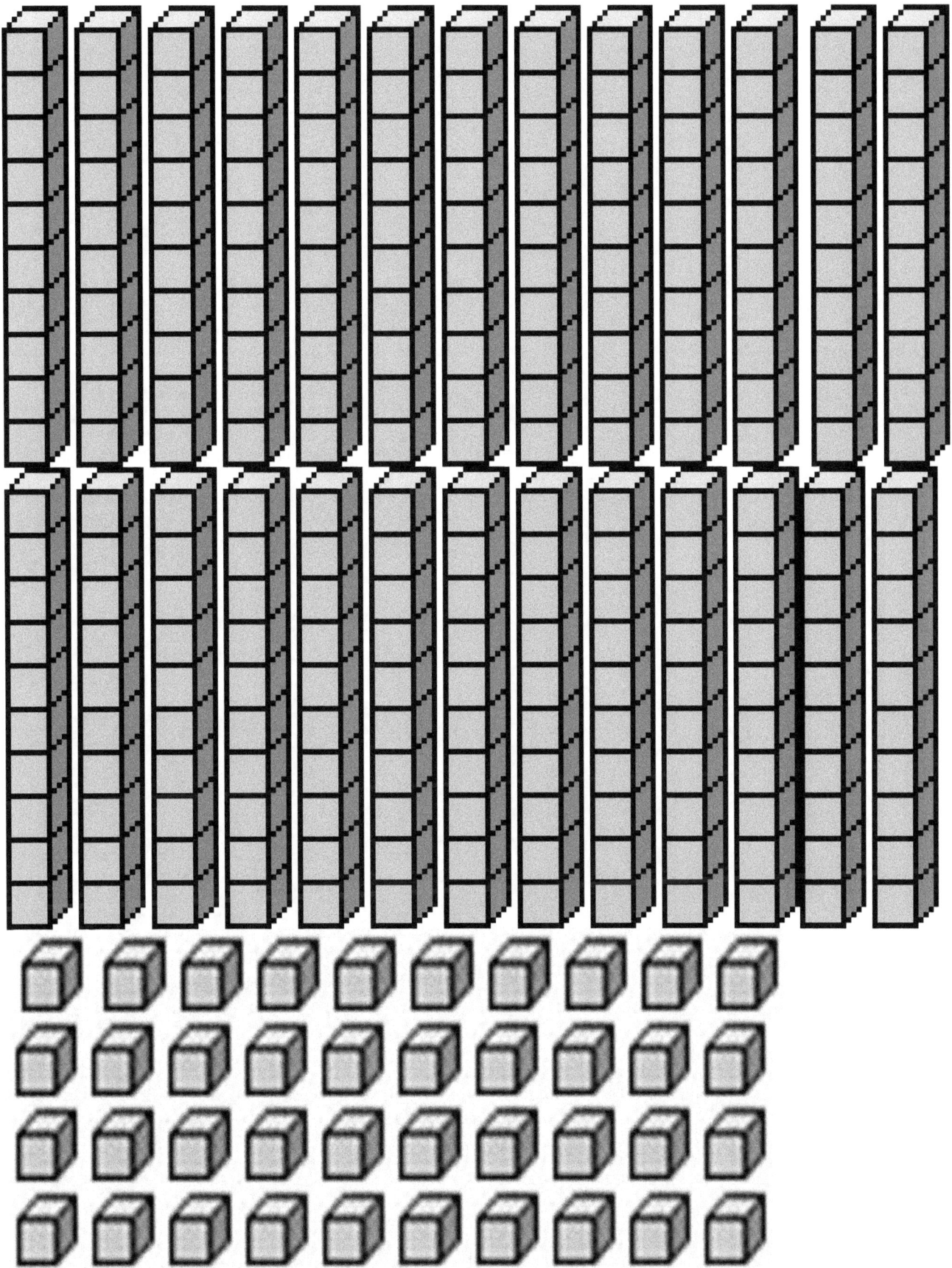

Name: _____

Use your math tiles to help you solve each problem.

1.
$$67$$
$$-28$$

2.
$$42$$
$$-34$$

3.
$$65$$
$$-47$$

4.
$$42$$
$$-36$$

5.
$$54$$
$$-29$$

6.
$$71$$
$$-27$$

7.
$$83$$
$$-25$$

8.
$$64$$
$$-18$$

9.
$$85$$
$$-36$$

10.
$$57$$
$$-38$$

Name: _____

Use your math tiles to help you solve each problem.

1. 96
   -48

2. 58
   - 29

3. 37
   -18

4. 56
   - 37

5. 65
   -26

6. 87
   -39

7. 92
   -47

8. 78
   -29

9. 25
   -16

10. 47
    -19

Name: _____

Use your math tiles to help you solve each problem.

1.   44
   -26

2.   36
   - 28

3.   17
   - 8

4.   65
   - 36

5.   41
   -13

6.   31
   -26

7.   44
   -35

8.   91
   -15

9.   55
   -46

10.   37
   -29

Name: _____

Use your math tiles to help you solve each problem.

1. 60
   -41

2. 73
   -35

3. 91
   -85

4. 61
   -27

5. 81
   -34

6. 50
   -33

7. 37
   -29

8. 41
   -22

9. 63
   -58

10. 41
   -35

Name: _____

Use your math tiles to help you solve each problem.

1. 66
    -47
    ____

2. 51
    - 32
    ____

3. 92
    -83
    ____

4. 54
    - 46
    ____

5. 83
    -55
    ____

6. 60
    -46
    ____

7. 35
    -27
    ____

8. 82
    -34
    ____

9. 71
    -64
    ____

10. 58
    -49
    ____

Name: _____

Use your math tiles to help you solve each problem.

1. 476
    − 67
    _____

2. 243
    − 39
    _____

3. 181
    − 78
    _____

4. 592
    − 56
    _____

5. 654
    − 26
    _____

6. 973
    − 15
    _____

7. 881
    − 45
    _____

8. 684
    − 38
    _____

9. 767
    − 29
    _____

10. 293
    − 14
    _____

Name: _____

Use your math tiles to help you solve each problem.

1. 562
    - 57

2. 381
    - 34

3. 271
    - 57

4. 452
    - 36

5. 541
    - 22

6. 864
    - 45

7. 796
    - 48

8. 632
    - 24

9. 551
    - 28

10. 378
    - 49

Name: _____

Use your math tiles to help you solve each problem.

1.   385
    − 26

2.   451
    − 49

3.   247
    − 38

4.   794
    − 85

5.   535
    − 16

6.   938
    − 19

7.   851
    − 45

8.   623
    − 18

9.   341
    − 29

10.   471
    − 29

Name: _____

Use your math tiles to help you solve each problem.

1. 741
   - 232

2. 472
   - 153

3. 291
   - 128

4. 643
   - 317

5. 544
   - 235

6. 972
   - 135

7. 860
   - 213

8. 361
   - 235

9. 250
   - 211

10. 482
    - 373

Name: _____

Use your math tiles to help you solve each problem.

1.   573
    - 275

2.   644
    - 535

3.   821
    - 318

4.   567
    - 418

5.   370
    - 129

6.   384
    - 235

7.   752
    - 645

8.   143
    - 127

9.   941
    - 736

10.   632
     - 414

Unit 13 – Borrowing in Subtraction

Borrowing larger numbers works the same ways a borrowing from the 10's to make more units.

```
  8   12
  X̶  2  6
- 7  6  4
_____
```

Ten 10 bars = one 100 block

```
  2   14
  X̶  4  8  9
- 1  6  3  5
_____
```

Ten 100 blocks = one 1,000 cube

```
  1   17
  X̶  7  6  4  8
- 1  9  3  3  4
_____
```

Ten 1,000 cubes = one 10,000's

1. 457
    - 65

2. 329
    - 39

3. 278
    - 87

4. 526
    - 92

Name: _____

1. 459
    -286

2. 917
    -746

3. 838
    -675

4. 629
    -449

5. 657
    -276

6. 715
    -275

7. 818
    -134

8. 829
    -166

9. 729
    -267

10. 645
    -173

Name: _____

1. 559
   -479

2. 648
   - 395

3. 539
   -176

4. 549
   -372

5. 758
   -294

6. 679
   -184

7. 549
   -368

8. 867
   -573

9. 658
   -267

10. 738
    -356

Name: _____

1. 7,459
   - 758

2. 4,259
   - 648

3. 5,178
   - 356

4. 2,567
   - 926

5. 9,577
   - 674

6. 3,277
   - 963

7. 2,548
   - 845

8. 8,369
   - 664

9. 6,678
   - 724

10. 7,189
    - 253

Name: _____

1. 4,450
   -1,750

2. 7,237
   - 6,721

3. 9,854
   -5,664

4. 4,545
   -2,824

5. 9,653
   -7,952

6. 6,138
   -5,716

7. 9,854
   -4,249

8. 3,169
   -1,526

9. 6,729
   -2,260

10. 4,357
    -1,645

Name: _____

Sometimes you will need to borrow more than one time to solve a problem. You will complete each column using the borrowing skills you have learned before moving onto the next column. Treat each borrowing action as a separate problem.

```
              2   15
        5  14         5   14
   9  3  X  4  3    9  X  X  4  3
-  3  1  8  5  2   -  3  1  8  5  2
_____   _____
            9  1    6  1  7  9  1
```

1. 5,164
    - 248

2. 9,381
    - 754

3. 3,161
    - 342

4. 6,295
    - 479

5. 7,583
    - 676

6. 4,193
    - 876

Name: _____

1. 30,462
   − 7,253

2. 56,291
   − 4,479

3. 24,939
   − 6,741

4. 81,574
   − 9,654

5. 63,657
   − 8,664

6. 20,735
   − 7,452

7. 51,844
   − 3,653

8. 42,165
   − 9,844

9. 76,729
   − 8,363

10. 24,353
    − 7,249

Name: _____

1.   76,529
    -12,748

2.   48,309
    -37,751

3.   57,129
    -23,643

4.   71,378
    -19,653

5.   82,624
    -16,345

6.   69,564
    -30,857

7.   83,545
    -56,562

8.   46,381
    -43,736

9.   42,518
    -19,455

10.   73,453
    -16,657

Name: _____

1. 160,754
    - 39,546

2. 455,217
    - 25,941

3. 714,238
    - 73,834

4. 316,524
    - 86,447

5. 563,335
    - 38,657

6. 220,775
    - 69,443

7. 655,844
    - 49,174

8. 808,195
    - 57,826

9. 376,729
    - 28,260

10. 534,359
    - 82,645

Name: _____

1. 502,479
    -472,548

2. 858,261
    -151,758

3. 753,937
    -155,640

4. 617,718
    -347,653

5. 780,629
    -209,845

6. 625,567
    -169,234

7. 457,433
    -366,562

8. 846,256
    - 353,746

9. 559,518
    -378,349

10. 672,742
    -206,657

144

Name: _____

1.   5,945,459
    - 869,536

2.   7,655,517
    - 570,744

3.   3,784,238
    - 273,639

4.   1,381,584
    - 656,447

5.   4,523,653
    - 451,334

6.   8,279,775
    - 569,228

7.   2,054,844
    - 145,834

8.   6,279,164
    - 557,836

9.   4,276,729
    - 328,262

10.   1,934,359
     - 529,645

Unit 14 – Breaking Down a Problem

Breaking a problem down into smaller parts can help to make subtracting large numbers easier. We call this breaking down a problem or using partial sums to subtract.

You can subtract 6,783 - 2,262 using the math skills that you have already learned or you can break the numbers down into smaller parts to make solving the problem easier.

$$6,783$$
$$-2,262$$

First, you will decompose the number into thousands, hundreds, tens, and ones.

	Thousands	Hundreds	Tens	Ones
	6,000	700	80	3
-	2,000	200	60	2

Now subtract each column starting with the ones and moving to the left toward the thousands.

	Thousands	Hundreds	Tens	Ones
	6,000	700	80	3
-	2,000	200	60	2
	4,000	500	20	1

Now, you can put the numbers back together to get the solution to the problem

$$6,783$$
$$-2,262$$
$$4,521$$

Name: _____

Decompose each number to help you solve the problem.

1.
    37,849
   - 37,140

Ten Thousands	Thousands	Hundreds	Tens	Ones

2.
    67,558
   - 11,211

Ten Thousands	Thousands	Hundreds	Tens	Ones

3.
    46,637
   - 23,120

Ten Thousands	Thousands	Hundreds	Tens	Ones

4.
    93,428
   - 2,201

Ten Thousands	Thousands	Hundreds	Tens	Ones

Name: _____

Decompose each number to help you solve the problem.

1.
34,452
- 2,361

Ten Thousands	Thousands	Hundreds	Tens	Ones

2.
27,512
- 1,435

Ten Thousands	Thousands	Hundreds	Tens	Ones

3.
61,282
- 3,617

Ten Thousands	Thousands	Hundreds	Tens	Ones

4.
42,324
-16,164

Ten Thousands	Thousands	Hundreds	Tens	Ones

Name: _____

Decompose each number to help you solve the problem.

1.
   45,563
   - 3,373

Ten Thousands	Thousands	Hundreds	Tens	Ones

2.
   38,323
   - 2,546

Ten Thousands	Thousands	Hundreds	Tens	Ones

3.
   72,393
   - 4,728

Ten Thousands	Thousands	Hundreds	Tens	Ones

4.
   53,435
   - 27,275

Ten Thousands	Thousands	Hundreds	Tens	Ones

Name: _____

Decompose each number to help you solve the problem.

1.

23,341

- 1,250

Ten Thousands	Thousands	Hundreds	Tens	Ones

2.

28,101

- 1,546

Ten Thousands	Thousands	Hundreds	Tens	Ones

3.

72,393

- 4,605

Ten Thousands	Thousands	Hundreds	Tens	Ones

4.

53,435

-26,264

Ten Thousands	Thousands	Hundreds	Tens	Ones

Name: _____

Decompose each number to help you solve the problem.

1.
	Ten Thousands	Thousands	Hundreds	Tens	Ones
45,563
- 3,326

2.
	Ten Thousands	Thousands	Hundreds	Tens	Ones
38,232
- 1,435

3.
	Ten Thousands	Thousands	Hundreds	Tens	Ones
72,373
- 4,606

4.
	Ten Thousands	Thousands	Hundreds	Tens	Ones
53,435
- 26,264

Unit 15 - Rounding

Another way to make solving difficult problems easier is rounding. Rounding is a way of simplifying numbers to make them easier to understand.

You can use rounding whenever you do not need to know the exact answer. Solving a problem without the exact answer is called estimating. You can estimate whenever the problem you are solving can be answered with a close number instead of the exact number.

Example: If you are feeding 48 peanuts to the squirrel in your yard, you do not need to know the exact number of peanuts you give to him. You can say that you gave the squirrel 50 peanuts and still be accurate. This is called an estimate of the number of peanuts you fed to the squirrel.

How do you know that 48 peanuts rounds to 50?

You will use place value when you are rounding numbers.   First, you need to find the round off digit. We are rounding the number of sticks to the nearest 10 so the number in the 10's place is the rounding number.

Tens   Ones

<u>**4**</u>   8

Next, we look at the number that is immediately to the right of the rounding number. That is the round off digit. In this example, the round off digit is the 8.

Tens   **Ones**

4   <u>**8**</u>

Next, you need to decide what to do with the 8. We round numbers to the nearest 10's place. 5 is the middle of the number line. If a number is 5 or more it is closer to the 10's so we round up. If a number is less than 5 it is closer to the 0 so we round down.

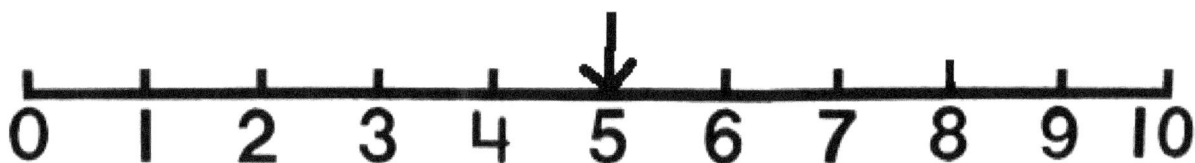

We need to round 48 to the nearest 10 place. 8 is our rounding off number.

Tens   **Ones**

4   <u>8</u>

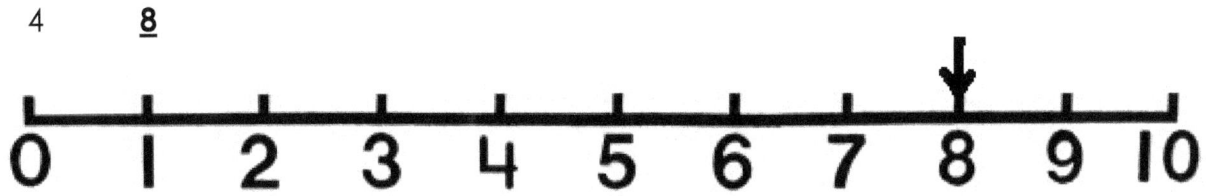

Now you will decide if the 8 is closer to the 10 or the 0.

8 is closer to the 10 so we will round the number 8 up to the next 10.

Our number was 48 so rounding up to the next 10 makes our rounded number 50.

Rounding to 10

Use the number line to practice rounding these numbers to the nearest 10.

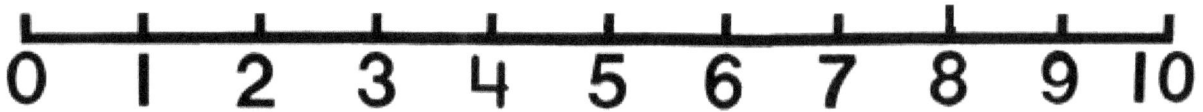

```
├──┼──┼──┼──┼──┼──┼──┼──┼──┼──┤
0   1   2   3   4   5   6   7   8   9   10
```

1.   Tens   Ones

    29

_____

2.   Tens   Ones

    51

_____

3.   Tens   Ones

    43

_____

4.   Tens   Ones

    18

_____

5.   Tens   Ones

    24

_____

6.   Tens   Ones

    35

_____-

_____

Name: _____

Use the number line to practice rounding these numbers to the nearest 10.

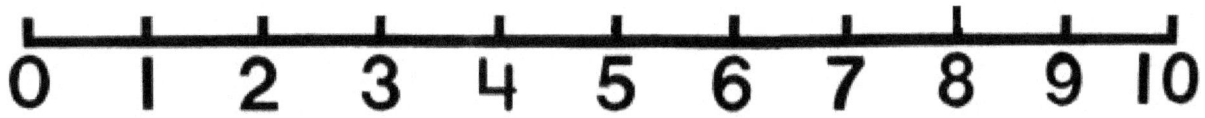

```
┠──┸──┸──┸──┸──┸──┸──┸──┸──┸──┨
0   1   2   3   4   5   6   7   8   9   10
```

1.   Tens   Ones
   57

   _____

2.   Tens   Ones
   32

   _____

3.   Tens   Ones
   78

   _____

4.   Tens   Ones
   16

   _____

5.   Tens   Ones
   91

   _____

6.   Tens   Ones
   29

   _____

Subtracting by Rounding to 10

A number line can also help you to add numbers when you do not need to know the exact answer.

If you have a subtraction problem that does not need an exact answer, you can round both numbers before subtracting.

Example: You buy 1 pound of peanuts to feed to the squirrels. There are about 175 peanuts in a pound. You already fed the squirrels 48 peanuts. About how many peanuts do you have left?

"About" is a clue word that tells you that you can answer the problem with an estimate instead of an exact calculation.

First, round each number to the nearest 10.

175 rounds up to 180 because the 5 is closer to 10 than to 0.

48 rounds up to 50 because the 8 is closer to 10 than to 0.

	1	7	5	Rounds To		1	8	0
		4	8	Rounds To	-		5	0
-								

Now subtract the numbers

	1	7	5	Rounds To		1	8	0
		4	8	Rounds To	-		5	0
-								
						1	3	0

You have about 130 peanuts left.

Name: _____

Use the number line to help you round and solve each problem

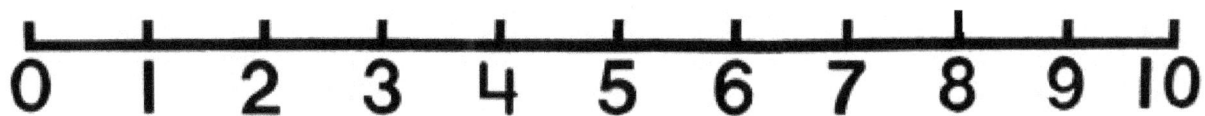

```
 ├──┼──┼──┼──┼──┼──┼──┼──┼──┼──┤
 0  1  2  3  4  5  6  7  8  9 10
```

1.   46 Rounds To
   - 32 Rounds To   - ____

2.   71 Rounds To
   - 18 Rounds To   - ____

3.   64 Rounds To
   - 35 Rounds To   - ____

4.   96 Rounds To
   - 22 Rounds To   - ____

5.   57 Rounds To
   - 32 Rounds To   - ____

6.   63 Rounds To
   - 34 Rounds To   - ____

7.   28 Rounds To
   - 21 Rounds To   - ____

8.   85 Rounds To
   - 12 Rounds To   - ____

Name: _____

Use the number line to help you round and solve each problem

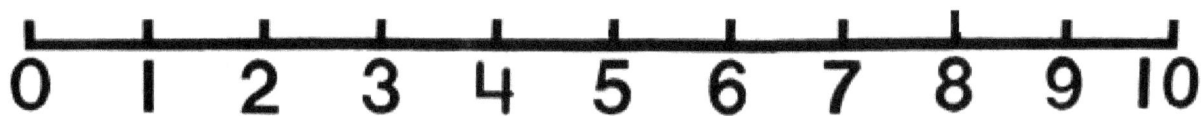

0 1 2 3 4 5 6 7 8 9 10

1. 67 Rounds To
   - 21 Rounds To   - _____

2. 82 Rounds To
   - 14 Rounds To   - _____

3. 45 Rounds To
   - 24 Rounds To   - _____

4. 27 Rounds To
   - 11 Rounds To   - _____

5. 27 Rounds To
   - 28 Rounds To   - _____

6. 74 Rounds To
   - 23 Rounds To   - _____

7. 39 Rounds To
   - 41 Rounds To   - _____

8. 54 Rounds To
   - 22 Rounds To   - _____

Name: _____

Use the number line to help you round and solve each problem

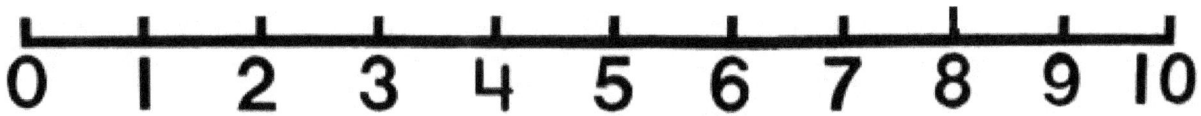

```
  ├──┼──┼──┼──┼──┼──┼──┼──┼──┼──┤
  0  1  2  3  4  5  6  7  8  9  10
```

1.   81   Rounds To
   - 12   Rounds To   - ____

2.   73   Rounds To
   - 24   Rounds To   - ____

3.   56   Rounds To
   - 33   Rounds To   - ____

4.   38   Rounds To
   - 22   Rounds To   - ____

5.   48   Rounds To
   - 37   Rounds To   - ____

6.   85   Rounds To
   - 14   Rounds To   - ____

7.   58   Rounds To
   - 47   Rounds To   - ____

8.   65   Rounds To
   - 25   Rounds To   - ____

Rounding to 100

Rounding becomes more important when you are working with larger numbers. Calculating 754 - 235 is a problem that you have the skills to solve. If you do not need to know the exact answer then you can make solving the problem easier by rounding.

You can round larger numbers using the same processes you learned when rounding tens and units.

First, decide which number is the rounding number.

You will always round the number immediately to the right of the rounding number.

Rounding 176 to the nearest 10 means that the 7 is the rounding number.

176		
hundreds	**tens**	ones
1	**7**	6

The number to the right of the 7 is a 6.

176		
hundreds	tens	**ones**
1	7	**6**

6 rounds up because it is closer to the 10 than to the 0 on the number line.

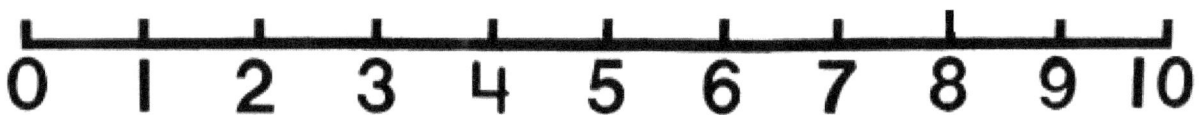

When we round up, we add the 10 that the 6 becomes to the rounding number making the answer 180.

Rounding 176 to the nearest 100 means that the 1 is the rounding number.

176

hundreds	tens	ones
_1_	7	6

The number to the right of the 1 is a 7.

176		
hundreds	**tens**	ones
1	_7_	6

7 rounds up because it is closer to the 10 than to the 0 on the number line.

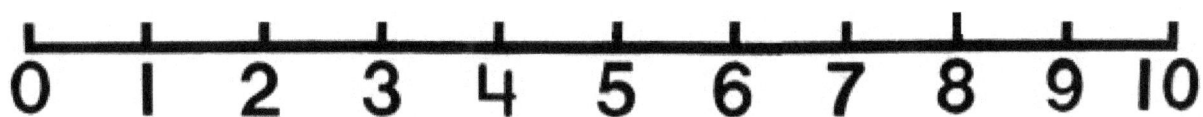

When we round up, we add the 100 that the 7 becomes to the rounding number making the answer 200.

Name: _____

Use the number line to practice rounding these numbers to the nearest 100.

```
├──┼──┼──┼──┼──┼──┼──┼──┼──┼──┤
0   1   2   3   4   5   6   7   8   9  10
```

1.   Hundreds Tens Ones   2.   Hundreds Tens Ones
   257   632

   _____   _____

3.   Hundreds Tens Ones   4.   Hundreds Tens Ones
                                     116
   578

   _____   _____

5.   Hundreds Tens Ones   6.   Hundreds Tens Ones
   391   829

   _____   _____

162

Name: _____

Use the number line to practice rounding these numbers to the nearest 100.

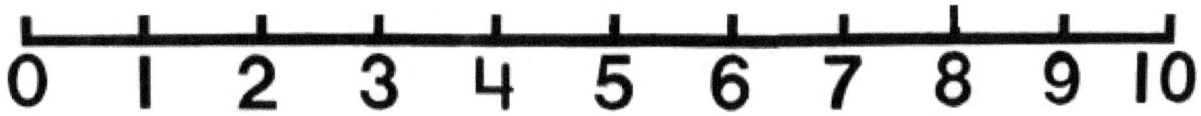

```
├──┼──┼──┼──┼──┼──┼──┼──┼──┼──┤
0   1   2   3   4   5   6   7   8   9  10
```

1.   Hundreds Tens Ones   2.   Hundreds Tens Ones

   368   275

   _____   _____

3.   Hundreds Tens Ones   4.   Hundreds Tens Ones

                                        584
   431

   _____   _____

5.   Hundreds Tens Ones   6.   Hundreds Tens Ones

   297   643

   _____   _____

Subtracting by Rounding to 100

A number line can also help you to subtract bigger numbers.

Example: Subtracting 675 - 237 is a problem that you have the skills to solve. If you do not need to now the exact answer and an estimate will do, then rounding can help to make solving the problem easier.

A lot of estimating problems will be story problems. When you see the word "about" it is a clue that you can use an estimate instead of calculating the exact answer.

First, round each number to the nearest 100.

6_7_5 rounds up to 700 because the 7 is closer to 10 than to 0.

2_3_7 rounds down to 200 because the 3 is closer to the 0 than to 10.

Remember that every number to the right of the one that you round turns into a 0.

$$
\begin{array}{r}
675 \\
-\ 237 \\
\end{array}
\quad
\begin{array}{l}
\text{Rounds To} \\
\text{Rounds To}
\end{array}
\quad
\begin{array}{r}
700 \\
-\ 200 \\
\end{array}
$$

Now solve the problem.

$$
\begin{array}{r}
675 \\
-\ 237 \\
\end{array}
\quad
\begin{array}{l}
\text{Rounds To} \\
\text{Rounds To}
\end{array}
\quad
\begin{array}{r}
700 \\
-\ 200 \\
\hline
500
\end{array}
$$

Subtracting 700 - 200 is easier than subtracting 675 - 237.

Name: _____

Use the number line to help you round and solve each problem

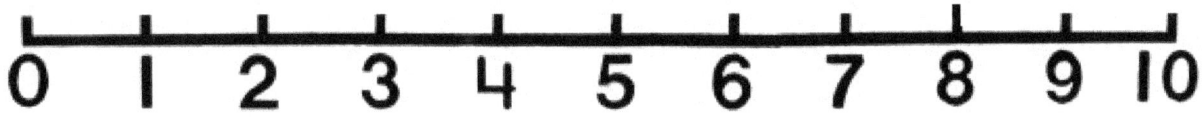

```
|___|___|___|___|___|___|___|___|___|___|
0   1   2   3   4   5   6   7   8   9  10
```

1.   746 Rounds To   2.   671 Rounds To

   - 532 Rounds To - _____   -118 Rounds To - _____

_____

3.   734 Rounds To   4.   816 Rounds To

    -235 Rounds To - _____   -472 Rounds To - _____

_____

5.   157 Rounds To   6.   563 Rounds To

   - 132 Rounds To - _____   -534 Rounds To - _____

_____

7.   528 Rounds To   8.   825 Rounds To

   - 341 Rounds To - _____   -112 Rounds To - _____

_____

165

Name: _____

Use the number line to help you round and solve each problem

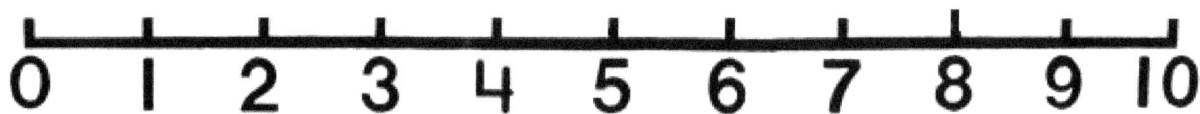

```
├──┴──┴──┴──┴──┴──┴──┴──┴──┴──┤
0   1   2   3   4   5   6   7   8   9   10
```

1.   746   Rounds To   2.   371   Rounds To

    - 232   Rounds To   -_____   -318
                                          Rounds To   -_____

---

3.   534   Rounds To   4.   816   Rounds To

    -335   Rounds To   -_____   -122   Rounds To   -_____

---

5.   757   Rounds To   6.   463   Rounds To

    -732   Rounds To   -_____   -434   Rounds To   -_____

---

7.   628   Rounds To   8.   285   Rounds To

    -341   Rounds To   -_____   -112   Rounds To   -_____

Name: _____

Use the number line to help you round and solve each problem

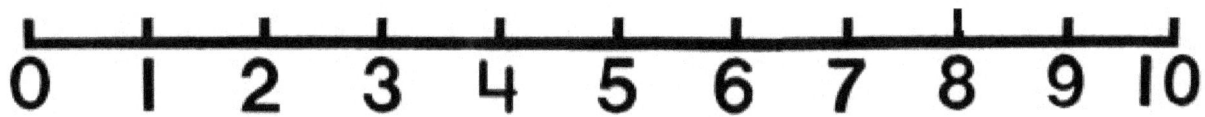

0 1 2 3 4 5 6 7 8 9 10

1.   721 Rounds To _____   2.   814 Rounds To _____

    - 267 Rounds To + _____   - 182 Rounds To + _____

3.   524 Rounds To _____   4.   611 Rounds To _____

    - 445 Rounds To + _____   - 327 Rounds To + _____

5.   627 Rounds To _____   6.   574 Rounds To _____

    - 328 Rounds To + _____   - 423 Rounds To + _____

7.   839 Rounds To _____   8.   722 Rounds To _____

    - 141 Rounds To + _____   - 324 Rounds To + _____

Name: _____

Use the number line to help you round and solve each problem

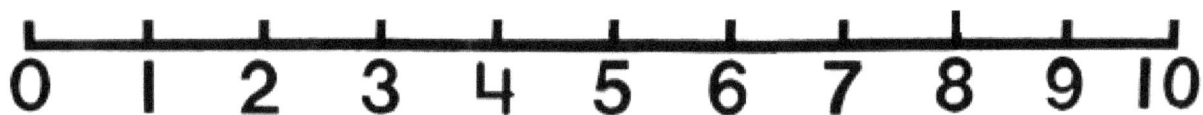

1.　612　Rounds To

　　- 381　Rounds To　+ _____

2.　873　Rounds To

　　- 124　Rounds To　+ _____

3.　556　Rounds To

　　- 433　Rounds To　+ _____

4.　722　Rounds To

　　- 238　Rounds To　+ _____

5.　748　Rounds To

　　- 237　Rounds To　+ _____

6.　485　Rounds To

　　- 514　Rounds To　+ _____

7.　858　Rounds To

　　- 147　Rounds To　+ _____

8.　665　Rounds To

　　- 325　Rounds To　+ _____

Rounding Bigger Numbers

Rounding becomes more important when you are working with larger numbers. 74,254 - 23,735 is a problem that you have the skills to solve, but if you do not need to know the exact answer then you can make solving the problem easier by rounding.

You can round larger numbers using the same processes you learned when rounding tens and units.

First, decide which number is the rounding number.

You will always round the number immediately to the right of the rounding number.

Rounding 74,254 to the nearest 100 means that the **2** is the rounding number.

74,254					
Ten Thousands	Thousands	,	**hundreds**	tens	ones
7	4	,	**2**	5	4

The number to the right of the 2 is a 5.

74,254					
Ten Thousands	Thousands	,	hundreds	**tens**	ones
7	4	,	2	**5**	4

5 rounds up because it is closer to the 10 than to the 0 on the number line.

When we round up, the 5 becomes a 0 and the 1 form the tens is added to the 100's column. Remember, every number to the right of the rounding number becomes a 0

74,254 rounded to the nearest 100 = 74,300

Rounding 74,254 to the nearest 1,000 means that the 4 is the rounding number.

74,254					
Ten Thousands	**Thousands**	,	hundreds	tens	ones
7	**4**	,	2	5	4

The number to the right of the 4 is a 2.

74,254					
Ten Thousands	Thousands	,	**hundreds**	tens	ones
7	4	,	**2**	5	4

2 rounds down because it is closer to the 0 than to the 10 on the number line.

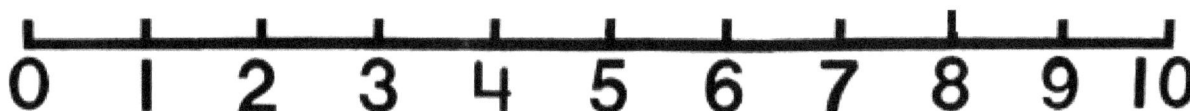

When we round down, we do not add anything to the thousands place because our number rounded to 0. Remember, every number to the right of the rounding number becomes a 0.

74,254 rounded to the nearest 1,000 = 74,000

Rounding 23,735 to the nearest 100 means that the 7 is the rounding number.

23,735					
Ten Thousands	Thousands	,	**hundreds**	tens	ones
2	3	,	**7**	3	5

The number to the right of the 7 is a 3.

23,735

Ten Thousands	Thousands	,	hundreds	**tens**	ones
2	3	,	7	**3**	5

3 rounds down because it is closer to the 0 than to the 10 on the number line.

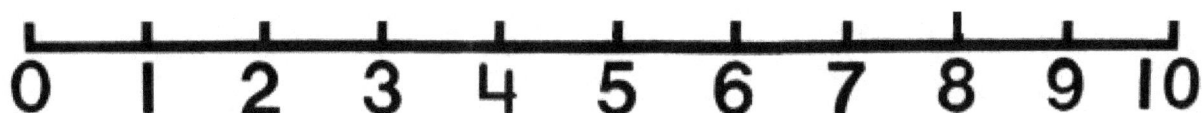

When we round down, the 3 becomes a 0 and we do not add anything to the 100 place. Remember, every number to the right of the rounding number becomes a 0

23,735 rounded to the nearest 100 = 23,700

Rounding 23,735 to the nearest 1,000 means that the 3 is the rounding number.

23,735					
Ten Thousands	**Thousands**	,	hundreds	tens	ones
2	**3**	,	7	3	5

The number to the right of the 3 is a 7.

23,735					
Ten Thousands	Thousands	,	**hundreds**	tens	ones
2	3	,	**7**	3	5

7 rounds up because it is closer to the 10 than to the 0 on the number line.

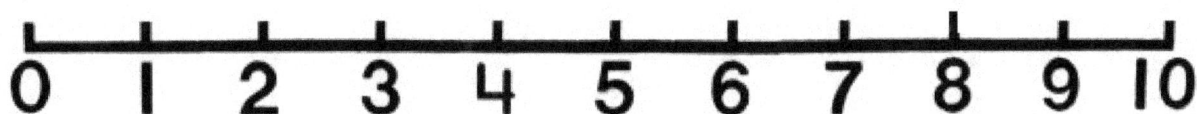

When we round up, we add the 1,000 that the 7 became to the thousands place. Remember, every number to the right of the rounding number becomes a 0.

23,735 rounded to the nearest 1,000 = 24,000

  74,254   Rounds To   74,000

-23,735   Rounds To   -24,000
                                     50,000

Name: _____

Use the number line to practice rounding these numbers to the nearest 1000.

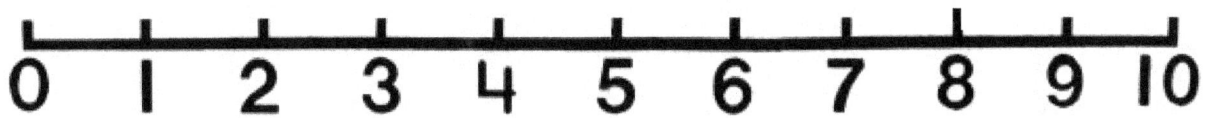

```
|--|--|--|--|--|--|--|--|--|--|
0  1  2  3  4  5  6  7  8  9  10
```

1.   Ten
   ThousandsThousandsHundredsTensOnes

   48,257

_____

2.   Ten
   ThousandsThousandsHundredsTensOnes

   71,864

_____

3.   Ten
   ThousandsThousandsHundredsTensOnes

   93,641

_____

4.   Ten
   ThousandsThousandsHundredsTensOnes

   23,864

_____

Name: _____

Use the number line to practice rounding these numbers to the nearest 1000.

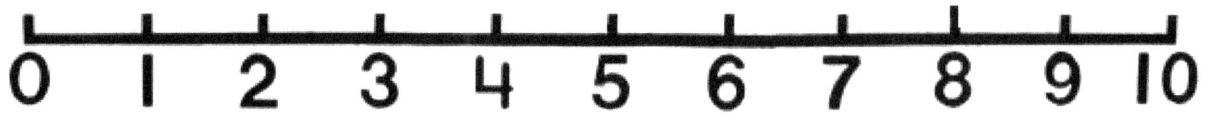

0   1   2   3   4   5   6   7   8   9   10

1.   Ten
              ThousandsThousandsHundredsTensOnes

   87,727

2.   Ten
              ThousandsThousandsHundredsTensOnes

   74,318

3.   Ten
              ThousandsThousandsHundredsTensOnes

   12,972

4.   Ten
              ThousandsThousandsHundredsTensOnes

   35,818

Name: _____

Use the number line to practice rounding these numbers to the nearest 1,000.

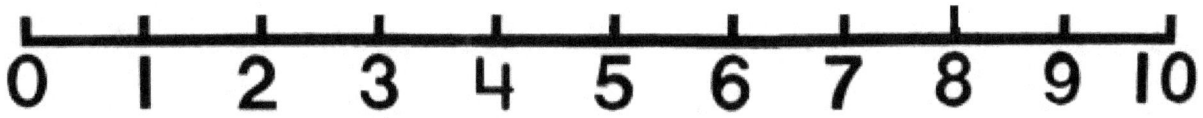

```
  |  |  |  |  |  |  |  |  |  |  |
  0  1  2  3  4  5  6  7  8  9  10
```

1.   Ten
          ThousandsThousandsHundredsTensOnes

   91,244

_____

2.   Ten
          ThousandsThousandsHundredsTensOnes

   18,635

_____

3.   Ten
          ThousandsThousandsHundredsTensOnes

   24,581

_____

4.   Ten
          ThousandsThousandsHundredsTensOnes

   46,427

_____

Name: _____

Use the number line to practice rounding these numbers to the nearest 1,000.

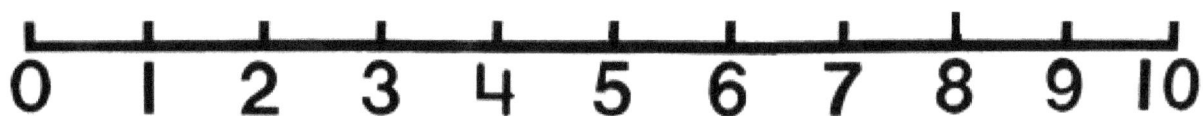

```
├──┼──┼──┼──┼──┼──┼──┼──┼──┼──┤
0   1   2   3   4   5   6   7   8   9  10
```

1.   Ten
        Thousands Thousands Hundreds Tens Ones

   35,818

_____

2.   Ten
        Thousands Thousands Hundreds Tens Ones

   87,727

_____

3.   Ten
        Thousands Thousands Hundreds Tens Ones

   74,318

_____

4.   Ten
        Thousands Thousands Hundreds Tens Ones

   94,611

_____

Name: _____

Use the number line to practice rounding these numbers to the nearest 10,000.

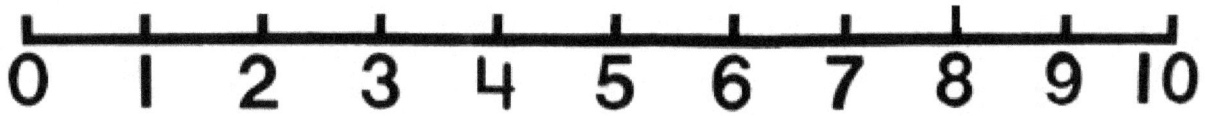

0   1   2   3   4   5   6   7   8   9   10

1.   Ten
            ThousandsThousandsHundredsTensOnes

    48,257

_____

2.   Ten
            ThousandsThousandsHundredsTensOnes

    71,864

_____

3.   Ten
            ThousandsThousandsHundredsTensOnes

    93,641

_____

4.   Ten
            ThousandsThousandsHundredsTensOnes

    17,916

_____

Name: _____

Use the number line to practice rounding these numbers to the nearest 10,000

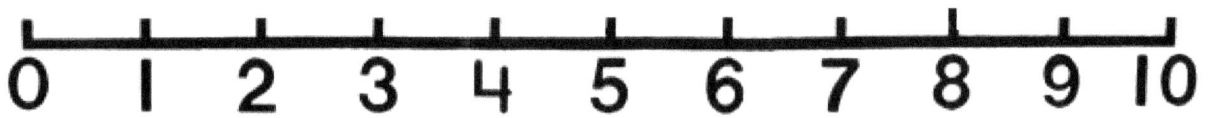

```
 |——|——|——|——|——|——|——|——|——|——|
 0  1  2  3  4  5  6  7  8  9  10
```

1.   Ten
         ThousandsThousandsHundredsTensOnes

  35,818

_____

2.   Ten
         ThousandsThousandsHundredsTensOnes

  87,727

_____

3.   Ten
         ThousandsThousandsHundredsTensOnes

  74,318

_____

4.   Ten
         ThousandsThousandsHundredsTensOnes

  23,144

_____

178

Name: _____

Use the number line to practice rounding these numbers to the nearest 100,000.

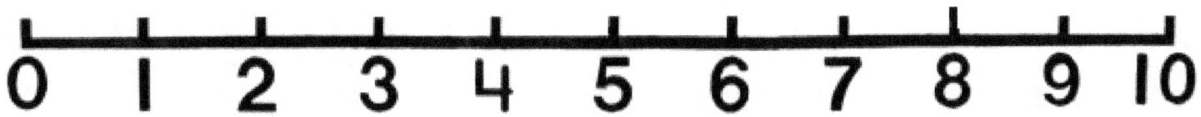

```
 L   I   I   I   I   I   I   I   I   I   L
 0   I   2   3   4   5   6   7   8   9  10
```

1.   Hundred   Ten
           ThousandsThousandsThousandsHundredsTensOnes

   391,244

2.   Hundred   Ten
           ThousandsThousandsThousandsHundredsTensOnes

   618,635

3.   Hundred   Ten
           ThousandsThousandsThousandsHundredsTensOnes

   824,581

4.   Hundred   Ten
           ThousandsThousandsThousandsHundredsTensOnes

   145,967

Name: _____

Use the number line to practice rounding these numbers to the nearest 100,000.

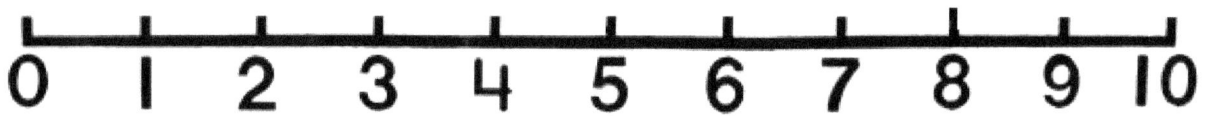

```
 ┠──┼──┼──┼──┼──┼──┼──┼──┼──┼──┨
 0  1  2  3  4  5  6  7  8  9 10
```

1.   Hundred   Ten
        ThousandsThousandsThousandsHundredsTensOnes

    435,818

2.   Hundred   Ten
        ThousandsThousandsThousandsHundredsTensOnes

    287,727

3.   Hundred   Ten
        ThousandsThousandsThousandsHundredsTensOnes

    574,318

4.   Hundred   Ten
        ThousandsThousandsThousandsHundredsTensOnes

    792,999

Name: _____

Use the number line to practice rounding these numbers to the nearest 1,000,000.

```
├──┴──┴──┴──┴──┴──┴──┴──┴──┴──┤
0   I   2   3   4   5   6   7   8   9  10
```

1.   Hundred   Ten
        Millions Thousands Thousands Thousands Hundreds Tens Ones

   7,391,244

_____

2.   Hundred   Ten
        Millions Thousands Thousands Thousands Hundreds Tens Ones

   4,618,635

_____

3.   Hundred   Ten
        Millions Thousands Thousands Thousands Hundreds Tens Ones

   5,824,581

_____

4.   Hundred   Ten
        Millions Thousands Thousands Thousands Hundreds Tens Ones

   2,986,555

_____

Name: _____

Use the number line to practice rounding these numbers to the nearest 1,000,000.

0 1 2 3 4 5 6 7 8 9 10

1.   Hundred   Ten
        Millions Thousands Thousands Thousands Hundreds Tens Ones

   9,435,818

2.   Hundred   Ten
        Millions Thousands Thousands Thousands Hundreds Tens Ones

   1,287,727

3.   Hundred   Ten
        Millions Thousands Thousands Thousands Hundreds Tens Ones

   3,574,318

4.   Hundred   Ten
        Millions Thousands Thousands Thousands Hundreds Tens Ones

   6,784,232

Subtracting Bigger Numbers by Rounding

A number line can also help you to subtract bigger numbers.

Example: Solving 3,213,464 - 1,675,231 is a problem that you have the skills to solve, but if you do not need to now the exact answer and an estimate will do, then rounding can help to make solving the problem easier.

First, round each number to the nearest 100,000.

3,213,464 rounds down to 3,200,000

1,675,231 rounds up to 1,700,000

Remember that every number to the right of the one that you round turns into a 0.

$$
\begin{array}{r}
3{,}213{,}464 \\
\underline{-1{,}675{,}231}
\end{array}
\quad
\begin{array}{l}
\text{Rounds To} \\
\text{Rounds To}
\end{array}
\quad
\begin{array}{r}
3{,}200{,}000 \\
\underline{-1{,}700{,}000}
\end{array}
$$

Now solve the problem.

$$
\begin{array}{r}
3{,}213{,}464 \\
\underline{-1{,}675{,}231}
\end{array}
\quad
\begin{array}{l}
\text{Rounds To} \\
\text{Rounds To}
\end{array}
\quad
\begin{array}{r}
3{,}200{,}000 \\
\underline{-1{,}700{,}000} \\
1{,}500{,}000
\end{array}
$$

Solving 3,200,000 - 1,700,000 is easier than solving 3,213,464 - 1,675,231.

Name: _____

Use the number line to help you round each number to the nearest 10,000 and solve each problem.

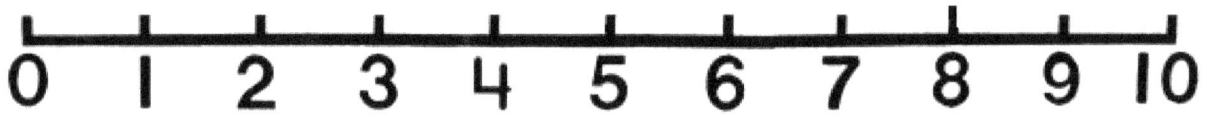

```
├──┼──┼──┼──┼──┼──┼──┼──┼──┼──┤
0   1   2   3   4   5   6   7   8   9  10
```

1.   24,446   Rounds To

    - 13,532   Rounds To   -_____

2.   51,734   Rounds To

    -36,235   Rounds To   -_____

3.   83,157   Rounds To

    -14,132   Rounds To   -_____

4.   45,528   Rounds To

    -41,341   Rounds To   -_____

Name: _____

Use the number line to help you round each number to the nearest 10,000 and solve each problem.

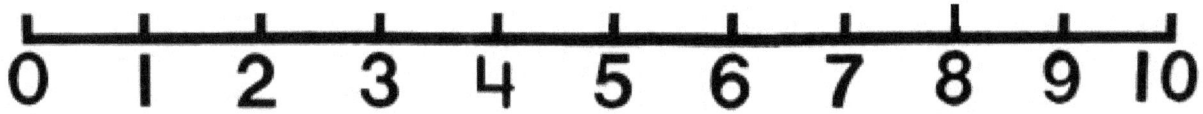

```
 ├──┼──┼──┼──┼──┼──┼──┼──┼──┼──┤
 0  1  2  3  4  5  6  7  8  9  10
```

1. 51,999   Rounds To

    - 26,318   Rounds To   - _____

2. 66,823   Rounds To

    -40,124   Rounds To   - _____

3. 72,046   Rounds To

    -13,021   Rounds To   - _____

4. 35,417   Rounds To

    -33,230   Rounds To   - _____

Name: _____

Use the number line to help you round each number to the nearest 100,000 and solve each problem.

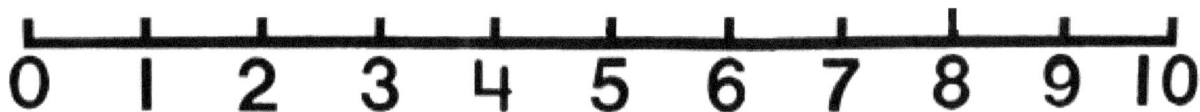

```
├──┼──┼──┼──┼──┼──┼──┼──┼──┼──┤
0   1   2   3   4   5   6   7   8   9  10
```

1.   637,523   Rounds To

      -122,518   Rounds To   - _____

2.   851,318   Rounds To

      -146,261   Rounds To   - _____

3.   534,921   Rounds To

      -318,032   Rounds To   - _____

4.   725,648   Rounds To

      -241,350   Rounds To   - _____

Name: _____

Use the number line to help you round each number to the nearest 100,000 and solve each problem.

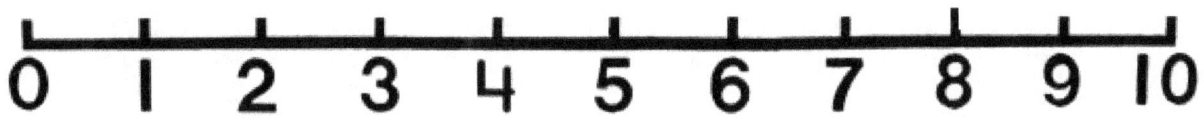

```
 |   |   |   |   |   |   |   |   |   |   |
 0   1   2   3   4   5   6   7   8   9  10
```

1. 634,876   Rounds To

    -251,123   Rounds To   - _____

2. 556,871   Rounds To

    -433,026   Rounds To   - _____

3. 872,369   Rounds To

    -123,520   Rounds To   - _____

4. 584,724   Rounds To

    -415,234   Rounds To   - _____

Name: _____

Use the number line to help you round each number to the nearest 1,000,000 and solve each problem.

```
|----|----|----|----|----|----|----|----|----|----|
0    1    2    3    4    5    6    7    8    9   10
```

1.   6,526,412   Rounds To
                   Rounds To
      -3,011,407
                         - _____

2.   7,035,207   Rounds To
                   Rounds To
      -2,740,150
                         - _____

3.   5,207,810   Rounds To
                   Rounds To
      - 4,423,021
                         - _____

4.   8,614,537   Rounds To
                   Rounds To
      - 1,130,240
                         - _____

Name: _____

Use the number line to help you round each number to the nearest 1,000,000 and solve each problem.

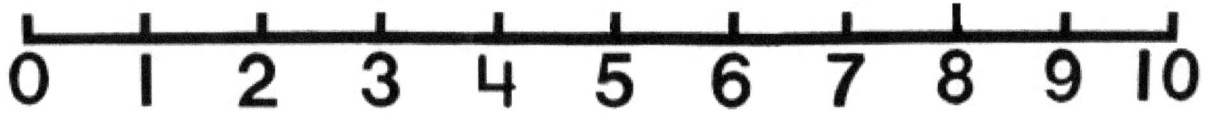

```
 ├──┼──┼──┼──┼──┼──┼──┼──┼──┼──┤
 0  I  2  3  4  5  6  7  8  9  10
```

1.   7,541,765   Rounds To

      -2,434,123   Rounds To   - _____

_____

2.   6,445,760   Rounds To

      -3,322,015   Rounds To   - _____

_____

3.   8,761,258   Rounds To

      -1,112,411   Rounds To   - _____

_____

4.   5,414,613   Rounds To

      - 4,584,285   Rounds To   - _____

_____

Unit 16 – Real World Story Problems

Sometimes you will solve math problems that look different from what you are used to. These math problems will use words to describe a real life situation. These are called story problems or real world problems. In life, you will use your math skills to solve many real world problems so it is important that you learn the steps that you must use to solve these types of problems.

You can use some simple steps to help make solving real world story problems easier.

- **Read the problem.**

    You must read the problem carefully to make certain that you understand what the problem is asking.

- **Ignore extra information.**

    Some story problems will contain information that is not important to solving the problem. You should read the problem to decide what is important and what is not important to helping you to solve the problem. Cross out any information that is not important so that it does not confuse you as you solve the problem.

- **Make a model of the problem.**

    Sometimes it is easiest to solve a story problem when you change the written information into pictures. Read the problem and decide if you can draw a picture to help you in solving the problem.

- **Think about the problem.**

    You should think about what the problem is asking you to solve.

- **Write the number sentence.**

    You should use all of the information that you decided is important to write a number sentence. The number sentence will be the actual problem.

- **Solve the problem.**

- **Check your work.**

You should compare your answer to what you think the problem is asking. Think about whether your answer makes sense when compared to the problem.

There are many steps to solving a story problem so you should always check your work to make sure that you did not make a simple calculation mistake.

Example: Sydney's cat had 7 kittens. Sydney gave one to her friend, two to her Grandmother and one to her cousin. Her teacher said she could not take one because she has a dog. How many kittens does Sydney have left?

Step 1:   Read the problem.

When you read this problem, you see the question "How many kittens does Sydney have left? This is the problem you are being asked to solve.

The word "left" is a clue word that tells us this is a subtraction problem.

Step 2:   Ignore any extra information.

The question is about cats so the parts of the question that tells us Sydney's teacher could not take a kitten because she has a dog is just extra information that does not help you to solve the problem.

When we remove the extra information, we are left with the information that helps us solve the problem.

Sydney's cat had 7 kittens. Sydney gave one to her friend, two to her Grandmother and one to her cousin.

Step 3:   Make a model of the problem.

Sydney has

Sydney's Friend   -

Sydney's Grandmother -

Sydney's Cousin

Step 4:   Think about the problem.

You need to decide what the problem is asking you to solve. This one is easy. The problem asks, "How many cats does Sydney have left?"

Step 5:   Write the number sentence.

7 – (1 – 2 – 1) =

Step 6:   Solve the problem.

7 – (1 – 2 – 1) = 3

Step 7:   Check your work.

When you check your work, you should think about the steps you took to solve the problem.

Does it make sense that Sydney has 3 kittens left?

The question asks have left. Have left usually tells us that we are working on a subtraction problem. Did you use subtraction to solve the problem?

Did you subtract correctly?

That problem was easy to solve. You will use the same steps to solve harder story problems.

Example: Stephen was selling flowers to raise money for his class trip. His teacher asked each student to sell 60 flowers. He sold 27 flowers to his neighbors, 8 flowers to his Dad's co-workers and 19 flowers to his family. How many more flowers does Stephen need to sell?

Step 1:   Read the problem.

When you read this problem, you see the question "How many more flowers does Stephen need to sell?" This is the problem that you are being asked to solve.

Step 2:   Ignore any extra information.

There is no extra information in this problem. Everything is important.

Step 3:   Make a model of the problem.

Goal   60

Neighbors   -27

+ Dad's Work   - 8

+ Family   - 19

Step 4:   Think about the problem.

You need to decide what the problem is asking you to solve. The problem asks, "How many more flowers does Stephen need to sell?"

Step 5:   Write the number sentence.

60 – 27 – 8 – 19 =   OR 60 – (27 + 8 + 19) =

Step 6:   Solve the problem.

There are 2 ways to solve this problem.

You can subtract each number from the goal.

```
  60
 -27
  33
 - 8
  25
 -19
   6 flowers left to sell
```

You can add all of the flowers that Stephen sold and then subtract them
from the goal.

```
  27            60
+  8           -54
 +19            6 flowers left to sell
  54
```

Step 7:   Check your work.

When you check your work, you should think about the steps you took to
solve the problem.

Does it make sense that the Stephen sold 54 flowers? Does it make sense
that Stephen has 6 flowers left to sell?

The question asks have left. Have left usually tells us that we are working
on a subtraction problem. Did you use subtraction to solve the problem?

Did you calculate correctly?

Name: _____

Follow the steps to solve each story problem.

1. The kids baked 12 cookies. Susan ate 5 cookies and Brian ate 3
   cookies. How many cookies are left?

Read the problem.
Ignore extra information.
Make a model of the problem.

Think about the problem.
Write the number sentence.

Solve the problem.
Check your work.

_____

2. Leslie had 18 trading cards. She gave 4 trading cards to her brother.
   How many trading cards does Leslie have left?

Read the problem.
Ignore extra information.
Make a model of the problem.

Think about the problem.
Write the number sentence.

Solve the problem.
Check your work.

_____

Name: _____

Follow the steps to solve each story problem.

1. There are 9 flowers growing in the garden, I picked 4 flowers and some greenery to put on the table. How many flowers are left in the garden?

Read the problem.
Ignore extra information.
Make a model of the problem.

Think about the problem.
Write the number sentence.

Solve the problem.
Check your work.

_____

2. Sarah's box has 15 crayons. Bill's box has 12 crayons. How many more crayons does Sarah have than Bill?

Read the problem.
Ignore extra information.
Make a model of the problem.

Think about the problem.
Write the number sentence.

Solve the problem.
Check your work.

_____

Name: _____

Follow the steps to solve each story problem.

1. Karen scored 51 points in the game and Steve scored 37 points. How many more points did Karen score in the game?

Read the problem.
Ignore extra information.
Make a model of the problem.

Think about the problem.
Write the number sentence.

Solve the problem.
Check your work.

---

2. Susan's book report book has 100 pages. Susan read 38 pages of her book on Monday, 38 on Tuesday, and 2 pages on Wednesday. How many pages does Susan have left to read?

Read the problem.
Ignore extra information.
Make a model of the problem.

Think about the problem.
Write the number sentence.

Solve the problem.
Check your work.

Name: _____

Follow the steps to solve each story problem.

1. Lisa sold 118 tickets to the school show and her sister sold 96. How many more tickets did Lisa sell?

_____

2. Penny has 41 girls and 27 boys in her karate class. Judy has 36 boys and 19 girls in her karate class. How many more students are in Penny's class than in Judy's class?

_____

3. Ben and Rick collected 97 tin cans on Saturday and 131 tin cans on Sunday. How many more tin cans did they collect on Sunday than on Saturday?

Name: _____

Follow the steps to solve each story problem.

1. The library had a book sale. They have 48 children's books, 126 non-fiction books, and 235 fiction books. They sold 125 fiction books, no non-fiction books, and 12 children's books on the first day. How many books did the Library have left for the second day?

2. Karen scored 56 points in today's game. Steve scored 11 fewer points than Karen. How many points did Steve score?

3. The girls caught 11 trout, 16 bluegill and 7 catfish while we were on vacation. 9 of the fish were too small and had to be thrown back. How many fish did the girls have left?

Name: _____

Follow the steps to solve each story problem.

1. Mark delivered 41 newspapers, Sarah delivered 22 magazines, and Jose delivered 36 newspapers. How many more newspapers did Mark deliver than the other children did?

2. 117 people rode the roller coaster, 96 people ate at the concession stand, and 137 people rode the bumper cars. How many fewer people ate at the concession stand instead of riding rides?

3. Our club got 25 new machines and 37 new members last month. We got 18 new machines and 31 new members this month. How many more members did our club get than machines?

Name: _____

Follow the steps to solve each story problem.

1. Samantha had to learn 75 spelling words for the spelling bee on Friday. She memorized 12 on Sunday, 18 on Monday, and 16 on Tuesday. She did not learn any new words on Wednesday. How many more words does Samantha have to learn before the spelling bee?

_____

2. Our team won our game 27-14. It was 92 degrees outside so the coach took all 21 of us for ice cream. 7 of us had chocolate cones, 9 had vanilla cones, and the rest ate yogurt cups. How many team members ate yogurt cups?

_____

3. Our music department has 12 tubas, 19 violins, and 11 flutes. We have 47 music students. How many more instruments than students does the music department have right now?

ANSWERS

Page 3	Page 5	Page 7	Page 8	Page 9
1. 1	1. 6	1. 3	1. 2	1. 2
2. 3	2. 5	2. 2	2. 5	2. 3
3. 1	3. 4	3. 1	3. 6	3. 5
4. 2	4. 3	4. 4	4. 8	4. 0
5. 4	5. 7	5. 0	5. 10	5. 1
		6. 5		

Page 10	Page 12	Page 14	Page 16	Page 18
1. 3	1. 2	1. 1	1. 0	1. 2
2. 1	2. 4	2. 3	2. 2	2. 1
3. 0	3. 1	3. 0	3. 3	3. 4
4. 4	4. 6	4. 5	4. 4	4. 2
5. 2	5. 5	5. 2	5. 1	5. 0

Page 20	Page 22	Page 24	Page 26	Page 28
1. 1	1. 6	1. 3	1. 3	1. 0
2. 0	2. 3	2. 5	2. 4	2. 3
3. 3	3. 2	3. 1	3. 0	3. 1
4. 1	4. 1	4. 0	4. 1	4. 2
5. 5	5. 0	5. 2	5. 5	5. 5

Page 30	Page 32	Page 33	Page 34	Page 35
1. 0	1. 13	1. 14	1. 12	1. 6
2. 15	2. 0	2. 13	2. 4	2. 6
3. 5	3. 2	3. 21	3. 7	3. 8
4. 0	4. 3	4. 4	4. 0	4. 5
5. 17				

Page 36	Page 37	Page 38	Page 39	Page 40
1. 16	1. 17	1. 8	1. 13	1. 9
2. 10	2. 10	2. 6	2. 0	2. 15
3. 5	3. 18	3. 7	3. 14	3. 9
4. 13	4. 13	4. 3	4. 6	4. 18

Page 57	Page 58	Page 59	Page 60	Page 61
1. 9	1. 8	1. 7	1. 6	1. 5
2. 5	2. 4	2. 3	2. 2	2. 1
3. 7	3. 6	3. 5	3. 4	3. 3
4. 2	4. 1	4. 0	4. 7	4. 8
5. 4	5. 3	5. 2	5. 1	5. 0
6. 3	6. 2	6. 1	6. 0	6. 6
7. 6	7. 5	7. 4	7. 3	7. 2
8. 8	8. 7	8. 6	8. 5	8. 4

Page 62	Page 63	Page 64	Page 65	Page 66
1. 4	1. 3	1. 7	1. 1	1. 15
2. 9	2. 5	2. 3	2. 2	2. 9
3. 2	3. 8	3. 0	3. 4	3. 8
4. 7	4. 6	4. 5	4. 0	4. 9
5. 6	5. 4	5. 4	5. 3	5. 15
6. 5	6. 1	6. 1	6. 0	
7. 1	7. 0	7. 6	7. 2	
8. 3	8. 2	8. 2	8. 1	
9. 8	9. 7	9. 9	9. 1	
10. 0	10. 1	10. 10	10. 2	

Page 68	Page 69	Page 70	Page 71
1. 11	1. 6	1. 14	1. 8
2. 8	2. 13	2. 6	2. 18
3. 7	3. 4	3. 14	3. 7
4. 4	4. 10	4. 0	4. 3

202

Page 75	Page 76	Page 77	Page 78
1. 64	1. 56	1. 38	1. 45
2. 53	2. 44	2. 2	2. 23
3. 13	3. 24	3. 76	3. 9
4. 19	4. 30	4. 45	4. 12
5. 9	5. 8	5. 5	5. 19
6. 34	6. 52	6. 8	6. 12
7. 24	7. 55	7. 34	7. 8
8. 53	8. 66	8. 64	8. 43
9. 16	9. 92	9. 82	9. 26
10. 34	10. 79	10. 35	10. 4

Page 79	Page 80	Page 81	Page 82	Page 83
1. 42	1. 45	1. 45	1. 24	1. 2
2. 10	2. 32	2. 31	2. 39	2. 45
3. 11	3. 24	3. 24	3. 24	3. 12
4. 32	4. 13	4. 44	4. 15	4. 55
5. 25	5. 64	5. 31	5. 41	5. 3
6. 45	6. 21	6. 23	6. 12	6. 34
7. 57	7. 43	7. 43	7. 43	7. 43
8. 46	8. 71	8. 51	8. 51	8. 55
9. 0	9. 14	9. 31	9. 52	9. 70
10. 13		10. 43	10. 73	10. 25
		11. 12	11. 19	11. 59
		12. 72	12. 73	12. 76

Page 84	Page 85	Page 86	Page 87	Page 88
1. 42	1. 152	1. 405	1. 124	1. 42
2. 37	2. 25	2. 346	2. 419	2. 495
3. 25	3. 214	3. 214	3. 494	3. 114
4. 37	4. 114	4. 524	4. 725	4. 135
5. 65	5. 231	5. 31	5. 251	5. 93
6. 30	6. 333	6. 394	6. 633	6. 534
7. 15	7. 343	7. 243	7. 343	7. 643
8. 21	8. 552	8. 596	8. 550	8. 514
9. 1	9. 772	9. 173	9. 512	9. 370
10. 47		10. 473	10. 153	10. 125
11. 21		11. 56	11. 319	11. 756
12. 3		12. 712	12. 74	12. 525

Page 89	Page 92	Page 93	Page 94	Page 95
1. 192	1. 9	1. 1	1. 11	1. 611
2. 527	2. 11	2. 4	2. 51	2. 251
3. 215	3. 11	3. 2	3. 21	3. 417
4. 157	4. 10	4. 5	4. 11	4. 725
5. 425	5. 11	5. 5	5. 31	5. 472
6. 10				
7. 755				
8. 274				
9. 401				
10. 247				
11. 559				
12. 524				

Page 96
1. 339
2. 374
3. 743
4. 115
5. 517

Page 97
1. 321
2. 364
3. 533
4. 105
5. 517

Page 99
1. 432
2. 219
3. 117
4. 504
5. 631
6. 965
7. 822
8. 630
9. 503
10. 223

Page 100
1. 323
2. 117
3. 105
4. 422
5. 523
6. 853
7. 710
8. 542
9. 611
10. 212

Page 101
1. 120
2. 206
3. 8
4. 111
5. 321
6. 552
7. 710
8. 143
9. 529
10. 111

Page 102
1. 220
2. 107
3. 9
4. 11
5. 521
6. 550
7. 220
8. 243
9. 417
10. 211

Page 103
1. 7,432
2. 4,219
3. 5,117
4. 1,544
5. 9,631
6. 3,965
7. 2,822
8. 8,632
9. 6,711
10. 7,221

Page 104
1. 5,321
2. 9,137
3. 3,105
4. 6,422
5. 7,521
6. 4,853
7. 1,710
8. 3,543
9. 2,656
10. 8,217

Page 105
1. 4,133
2. 7,216
3. 8,108
4. 1,121
5. 3,321
6. 5,552
7. 9,710
8. 2,040
9. 6,509
10. 4,013

Page 106
1. 2,217
2. 8,217
3. 5,359
4. 7,112
5. 1,321
6. 9,330
7. 3,134
8. 6,223
9. 9,237
10. 4,015

Page 107
1. 4,126
2. 7,516
3. 8,101
4. 1,101
5. 3,322
6. 5,552
7. 9,710
8. 2,443
9. 6,549
10. 4,312

Page 108
1. 3,403
2. 5,596
3. 4,518
4. 7,032
5. 5,213
6. 2,421
7. 6,140
8. 7,235
9. 2,329
10. 1,213

Page 109
1. 30,126
2. 51,536
3. 11,408
4. 85,121
5. 62,321
6. 22,552
7. 53,710
8. 45,443
9. 72,549
10. 22,312

Page 110
1. 60,317
2. 17,650
3. 32,629
4. 61,340
5. 79,322
6. 39,330
7. 23,130
8. 3,523
9. 37,237
10. 64,315

Page 111
1. 139,024
2. 431,536
3. 761,308
4. 375,121
5. 532,321
6. 242,552
7. 614,710
8. 815,743
9. 254,549
10. 510,312

Page 112
1. 170,317
2. 7,657
3. 642,629
4. 136,340
5. 583,321
6. 549,330
7. 122,134
8. 713,525
9. 269,237
10. 474,322

Page 113
1. 5,739,126
2. 7,331,516
3. 3,561,408
4. 1,374,121
5. 4,132,332
6. 8,349,552
7. 2,514,710
8. 6,715,741
9. 4,155,519
10. 1,122,212

Page 125
1. 39
2. 8
3. 18
4. 6
5. 25
6. 44
7. 58
8. 46
9. 49
10. 19

Page 126
1. 48
2. 29
3. 19
4. 19
5. 39
6. 48
7. 45
8. 49
9. 9
10. 28

Page 127
1. 18
2. 8
3. 9
4. 29
5. 28
6. 5
7. 9
8. 76
9. 9
10. 8

Page 128
1. 19
2. 38
3. 6
4. 34
5. 47
6. 17
7. 8
8. 19
9. 5
10. 6

Page 129
1. 19
2. 19
3. 9
4. 8
5. 28
6. 14
7. 8
8. 48
9. 7
10. 9

Page 130
1. 409
2. 204
3. 103
4. 536
5. 628
6. 958
7. 836
8. 646
9. 738
10. 279

Page 131
1. 505
2. 347
3. 214
4. 416
5. 519
6. 819
7. 748
8. 608
9. 523
10. 329

Page 132
1. 359
2. 402
3. 209
4. 709
5. 519
6. 919
7. 806
8. 605
9. 312
10. 442

Page 133
1. 509
2. 319
3. 163
4. 326
5. 309
6. 837
7. 647
8. 126
9. 39
10. 109

Page 134
1. 298
2. 109
3. 503
4. 149
5. 241
6. 149
7. 107
8. 16
9. 205
10. 218

Page 135
1. 327
2. 290
3. 191
4. 434

Page 136
1. 173
2. 171
3. 163
4. 180
5. 381
6. 440
7. 684
8. 663
9. 462
10. 472

Page 137
1. 80
2. 253
3. 363
4. 177
5. 464
6. 495
7. 181
8. 294
9. 391
10. 382

Page 138
1. 6,701
2. 3,611
3. 4,822
4. 1,641
5. 8,903
6. 2,314
7. 1,703
8. 7,705
9. 5,954
10. 6,936

Page 139
1. 2,700
2. 516
3. 4,190
4. 1,721
5. 1,701
6. 422
7. 5,605
8. 1,643
9. 4,469
10. 2,712

Page 140
1. 4,916
2. 8,627
3. 2,819
4. 5,816
5. 6,907
6. 3,317

Page 141
1. 23,209
2. 51,812
3. 18,198
4. 71,920
5. 54,993
6. 13,283
7. 48,191
8. 32,321
9. 68,366
10. 17,104

Page 142
1. 63,781
2. 10,558
3. 33,486
4. 51,725
5. 66,279
6. 38,707
7. 26,983
8. 2,645
9. 23.063
10. 56.796

Page 143
1. 121,208
2. 429,276
3. 640,404
4. 230,077
5. 524,677
6. 151,332
7. 606,670
8. 750,369
9. 348,469
10. 451,714

Page 144
1. 29,931
2. 706,503
3. 598,297
4. 270,065
5. 570,784
6. 456,333
7. 90,871
8. 492,510
9. 181,169
10. 466,085

Page 145
1. 5,075,923
2. 7,084,773
3. 3,510,599
4. 725,137
5. 4,072,319
6. 7,710,547
7. 1,763,176
8. 5,721,328
9. 3,948,467
10. 1,404,714

Page 147
1. 709
2. 56,347
3. 23,517
4. 91,227

Page 148
1. 32,091
2. 26,077
3. 57,665
4. 26,160

Page 149
1. 42,190
2. 35,777
3. 67,665
4. 26,160

Page 150
1. 22,091
2. 26,555
3. 67,788
4. 27,171

Page 151
1. 42,237
2. 36,797
3. 67,767
4. 27,171

Page 154
1. 30
2. 50
3. 40
4. 20
5. 20
6. 40

Page 155
1. 60
2. 30
3. 80
4. 20
5. 90
6. 30

Page 157
1. 50
   30
   20
2. 70
   20
   50
3. 60
   40
   20
4. 100
   20
   80

5. 60
   30
   30
6. 60
   30
   30
7. 30
   20
   10
8. 90
   10
   80

Page 158
1. 70
   20
   50
2. 80
   10
   70
3. 50
   20
   30
4. 30
   10
   20

5. 30
   30
   0
6. 70
   20
   50
7. 40
   40
   0
8. 50
   20
   30

Page 159

1. 80
   10
   70
2. 70
   20
   50
3. 60
   30
   40
4. 40
   20
   20

5. 50
   40
   10
6. 90
   10
   80
7. 60
   50
   10
8. 70
   30
   40

Page 162

1. 300
2. 600
3. 600
4. 100
5. 400
6. 800

Page 163

1. 400
2. 300
3. 400
4. 600
5. 300
6. 600

Page 165

1. 700
   500
   200
2. 700
   100
   600
3. 700
   200
   500
4. 800
   500
   300

5. 200
   100
   100
6. 600
   500
   100
7. 500
   300
   200
8. 800
   100
   700

Page 166

1. 700
   200
   500
2. 400
   300
   100
3. 500
   300
   200
4. 800
   100
   700

5. 800
   700
   300
6. 500
   400
   100
7. 600
   300
   300
8. 200
   100
   100

Page 167

1. 700
   300
   400
2. 800
   200
   600
3. 500
   400
   100
4. 600
   300
   300

5. 600
   300
   300
6. 600
   400
   200
7. 800
   100
   700
8. 700
   300
   400

Page 168

1. 600
   400
   200
2. 900
   100
   800
3. 600
   400
   200
4. 700
   200
   500

5. 700
   200
   500
6. 500
   500
   100
7. 900
   100
   800
8. 700
   300
   400

Page 173

1. 48,000
2. 72,000
3. 94,000
4. 24,000

Page 174

1. 88,000
2. 74,000
3. 13,000
4. 36,000

Page 175

1. 91,000
2. 19,000
3. 25,000
4. 46,000

Page 176

1. 36,000
2. 88,000
3. 74,000
4. 95,000

Page 177

1. 50,000
2. 70,000
3. 90,000
4. 20,000

Page 178

1. 40,000
2. 90,000
3. 70,000
4. 20,000

Page 179

1. 400,000
2. 600,000
3. 800,000
4. 100,000

Page 180

1. 7,000,000
2. 5,000,000
3. 6,000,000
4. 3,000,000

Page 182

1. 9,000,000
2. 1,000,000
3. 4,000,000
4. 7,000,000

Page 184	Page 185	Page 186	Page 187	Page 188
1. 20,000	1. 50.000	1. 600.000	1. 600.000	1. 6.000.000
<u>10,000</u>	<u>30.000</u>	<u>100.000</u>	<u>300.000</u>	<u>3.000.000</u>
10,000	20.000	500.000	300.000	3.000.000
2. 50,000	2. 70.000	2. 900.000	2. 600.000	2. 7.000.000
<u>40,000</u>	<u>40.000</u>	<u>100.000</u>	<u>400.000</u>	<u>3.000.000</u>
10,000	30.000	80,000	200.000	4.000.000
3. 80,000	3. 70.000	3. 500.000	3. 900.000	3. 5.000.000
<u>10,000</u>	<u>10.000</u>	<u>300.000</u>	<u>100.000</u>	<u>4.000.000</u>
70.000	60.000	200.000	800.000	1.000.000
4. 50,000	4. 40.000	4. 700.000	4. 600.000	4. 9.000.000
<u>40,000</u>	<u>30.000</u>	<u>200.000</u>	<u>400.000</u>	<u>1.000.000</u>
10.000	10.000	500.000	200.000	8.000.000

Page 189
1. 8,000,000
   <u>2,000,000</u>
   6,000,000
2. 6,000,000
   <u>3,000,000</u>
   3,000,000
3. 9,000,000
   <u>1,000,000</u>
   8,000,000
4. 5,000,000
   <u>5,000,000</u>
          0

Page 195
1. $5 + 3 = 8$
   $12 - 8 = 4$
2. $18 - 4 = 14$

Page 196
1. $9 - 4 = 5$
2. $15 - 12 = 3$

Page 197
1. $51 - 37 = 14$
2. $38 + 38 + 2 = 78$
   $100 - 78 = 22$

Page 198
1. $118 - 96 = 22$
2. $41 + 27 = 68$
   $36 + 19 = 55$
   $68 - 55 = 13$
3. $131 - 97 = 34$

Page 199
1. $48 + 126 + 235 = 409$
   $125 + 12 = 137$
   $409 - 137 = 272$
2. $56 - 11 = 45$
3. $11 + 16 + 7 = 34$
   $34 - 9 = 25$

Page 200
1. $40 - 36 = 5$
2. $117 + 137 = 254$
   $234 - 96 = 138$
3. $25 + 18 = 43$
   $37 + 31 = 68$
   $68 - 43 = 25$

Page 201
1. $12 + 18 + 16 = 46$
   $75 - 46 = 29$
2. $7 + 9 = 16$
   $21 - 16 = 5$
3. $12 + 19 + 11 = 42$
   $47 - 42 = 5$

www.ingramcontent.com/pod-product-compliance
Lightning Source LLC
LaVergne TN
LVHW081333060426
835513LV00014B/1264